Boolean Differential Equations

Synthesis Lectures on Digital Circuits and Systems

Editor
Mitchell A. Thornton, *Southern Methodist University*

The Synthesis Lectures on Digital Circuits and Systems series is comprised of 50- to 100-page books targeted for audience members with a wide-ranging background. The Lectures include topics that are of interest to students, professionals, and researchers in the area of design and analysis of digital circuits and systems. Each Lecture is self-contained and focuses on the background information required to understand the subject matter and practical case studies that illustrate applications. The format of a Lecture is structured such that each will be devoted to a specific topic in digital circuits and systems rather than a larger overview of several topics such as that found in a comprehensive handbook. The Lectures cover both well-established areas as well as newly developed or emerging material in digital circuits and systems design and analysis.

Boolean Differential Equations
Bernd Steinbach and Christian Posthoff
2013

Bad to the Bone: Crafting Electronic Systems with BeagleBone and BeagleBone Black
Steven F. Barrett and Jason Kridner
2013

Introduction to Noise-Resilient Computing
S.N. Yanushkevich, S. Kasai, G. Tangim, A.H. Tran, T. Mohamed, and V.P. Smerko
2013

Atmel AVR Microcontroller Primer: Programming and Interfacing, Second Edition
Steven F. Barrett and Daniel J. Pack
2012

Representation of Multiple-Valued Logic Functions
Radomir S. Stankovic, Jaakko T. Astola, and Claudio Moraga
2012

Arduino Microcontroller: Processing for Everyone! Second Edition
Steven F. Barrett
2012

Advanced Circuit Simulation Using Multisim Workbench
David Báez-López, Félix E. Guerrero-Castro, and Ofelia Delfina Cervantes-Villagómez
2012

Circuit Analysis with Multisim
David Báez-López and Félix E. Guerrero-Castro
2011

Microcontroller Programming and Interfacing Texas Instruments MSP430, Part I
Steven F. Barrett and Daniel J. Pack
2011

Microcontroller Programming and Interfacing Texas Instruments MSP430, Part II
Steven F. Barrett and Daniel J. Pack
2011

Pragmatic Electrical Engineering: Systems and Instruments
William Eccles
2011

Pragmatic Electrical Engineering: Fundamentals
William Eccles
2011

Introduction to Embedded Systems: Using ANSI C and the Arduino Development Environment
David J. Russell
2010

Arduino Microcontroller: Processing for Everyone! Part II
Steven F. Barrett
2010

Arduino Microcontroller Processing for Everyone! Part I
Steven F. Barrett
2010

Digital System Verification: A Combined Formal Methods and Simulation Framework
Lun Li and Mitchell A. Thornton
2010

Progress in Applications of Boolean Functions
Tsutomu Sasao and Jon T. Butler
2010

Embedded Systems Design with the Atmel AVR Microcontroller: Part II
Steven F. Barrett
2009

Embedded Systems Design with the Atmel AVR Microcontroller: Part I
Steven F. Barrett
2009

Embedded Systems Interfacing for Engineers using the Freescale HCS08 Microcontroller II:
Digital and Analog Hardware Interfacing
Douglas H. Summerville
2009

Designing Asynchronous Circuits using NULL Convention Logic (NCL)
Scott C. Smith and JiaDi
2009

Embedded Systems Interfacing for Engineers using the Freescale HCS08 Microcontroller I:
Assembly Language Programming
Douglas H.Summerville
2009

Developing Embedded Software using DaVinci & OMAP Technology
B.I. (Raj) Pawate
2009

Mismatch and Noise in Modern IC Processes
Andrew Marshall
2009

Asynchronous Sequential Machine Design and Analysis: A Comprehensive Development of
the Design and Analysis of Clock-Independent State Machines and Systems
Richard F. Tinder
2009

An Introduction to Logic Circuit Testing
Parag K. Lala
2008

Pragmatic Power
William J. Eccles
2008

Multiple Valued Logic: Concepts and Representations
D. Michael Miller and Mitchell A. Thornton
2007

Finite State Machine Datapath Design, Optimization, and Implementation
Justin Davis and Robert Reese
2007

Atmel AVR Microcontroller Primer: Programming and Interfacing
Steven F. Barrett and Daniel J. Pack
2007

Pragmatic Logic
William J. Eccles
2007

PSpice for Filters and Transmission Lines
Paul Tobin
2007

PSpice for Digital Signal Processing
Paul Tobin
2007

PSpice for Analog Communications Engineering
Paul Tobin
2007

PSpice for Digital Communications Engineering
Paul Tobin
2007

PSpice for Circuit Theory and Electronic Devices
Paul Tobin
2007

Pragmatic Circuits: DC and Time Domain
William J. Eccles
2006

Pragmatic Circuits: Frequency Domain
William J. Eccles
2006

Pragmatic Circuits: Signals and Filters
William J. Eccles
2006

High-Speed Digital System Design
Justin Davis
2006

Introduction to Logic Synthesis using Verilog HDL
Robert B.Reese and Mitchell A.Thornton
2006

Microcontrollers Fundamentals for Engineers and Scientists
Steven F. Barrett and Daniel J. Pack
2006

Boolean Differential Equations

Bernd Steinbach and Christian Posthoff

ISBN:978-3-031-79860-3   paperback
ISBN:978-3-031-79861-0   ebook

DOI 10.1007/978-3-031-79861-0

A Publication in the Springer series
SYNTHESIS LECTURES ON DIGITAL CIRCUITS AND SYSTEMS

Lecture #42
Series Editor: Mitchell A. Thornton, *Southern Methodist University*
Series ISSN
Synthesis Lectures on Digital Circuits and Systems
Print 1932-3166   Electronic 1932-3174

Boolean Differential Equations

Bernd Steinbach
Freiberg University of Mining and Technology, Germany

Christian Posthoff
The University of The West Indies, Trinidad & Tobago

SYNTHESIS LECTURES ON DIGITAL CIRCUITS AND SYSTEMS #42

ABSTRACT

The Boolean Differential Calculus (BDC) is a very powerful theory that extends the structure of a Boolean Algebra significantly. Based on a small number of definitions, many theorems have been proven. The available operations have been efficiently implemented in several software packages. There is a very wide field of applications. While a Boolean Algebra is focused on values of logic functions, the BDC allows the evaluation of *changes* of function values. Such changes can be explored for pairs of function values as well as for whole subspaces. Due to the same basic data structures, the BDC can be applied to any task described by logic functions and equations together with the Boolean Algebra. The BDC can be widely used for the analysis, synthesis, and testing of digital circuits.

Generally speaking, a *Boolean differential equation* (BDE) is an equation in which elements of the BDC appear. It includes variables, functions, and derivative operations of these functions. The solution of such a BDE is a *set of Boolean functions*. This is a significant extension of Boolean equations, which have sets of Boolean vectors as solutions. In the simplest BDE a derivative operation of the BDC on the left-hand side is equal to a logic function on the right-hand side. The solution of such a simple BDE means to execute an operation which is inverse to the given derivative. BDEs can be applied in the same fields as the BDC, however, their possibility to express sets of Boolean functions extends the application field significantly.

KEYWORDS

Boolean differential equation, Boolean Differential Calculus, Boolean Algebra, Boolean function, Boolean equation, set of Boolean functions, XBOOLE

Contents

1 Basics of the Binary Boolean Algebra 1
   1.1   Introduction 1
   1.2   Boolean Space, Boolean Functions, and Operations 1
   1.3   Boolean Functions and Operations 2
   1.4   Boolean Equations 8
   1.5   Systems of Equations and Inequalities 9
   1.6   Solutions with Regard to Variables 10
   1.7   XBOOLE 15
       1.7.1 Concept and Properties... 15
       1.7.2 XBOOLE Monitor 16
       1.7.3 XBOOLE Problem Program 25
       1.7.4 XBOOLE Library 28

2 Summary of the Boolean Differential Calculus 31
   2.1   Introduction 31
   2.2   Simple Derivative Operations 32
   2.3   Vectorial Derivatives 37
   2.4 m-fold Derivative Operations 41

3 Boolean Differential Equations 49
   3.1   Introduction 49
   3.2   Boolean Differential Equations of a Single Simple Derivative 49
   3.3   Boolean Differential Equations of a Single Vectorial Derivative...... 52
   3.4   Boolean Differential Equations Restricted to All Vectorial Derivatives........ 56
       3.4.1 Essence of Boolean Differential Equations 56
       3.4.2 Classes of Solution Functions 64
       3.4.3 Separation of Function Classes 70
       3.4.4 Separation of Function Classes using XBOOLE 77
   3.5   Boolean Differential Equations for Each Set of Solution Functions 86
       3.5.1 Separation of Functions 86
       3.5.2 Separation of Functions using XBOOLE 97

3.6   Boolean Differential Equations of All Derivative Operations 105

    3.6.1   Examples 105

    3.6.2   Extension to All Simple Derivative Operations 111

    3.6.3   Extension to All Vectorial Derivative Operations 112

    3.6.4   Extension to All m-fold Derivative Operations 113

    3.6.5   Methods of Transformations as Preprocessing Step 115

3.7   Most General Boolean Differential Equations 122

4   Solutions of the Exercises **129**

4.1   Solution of Chapter 1 129

4.2   Solution of Chapter 2 130

4.3   Solution of Chapter 3 131

Bibliography **139**

Authors' Biographies **143**

Index **145**

CHAPTER 1

Basics of the Binary Boolean Algebra

1.1 INTRODUCTION

We start this chapter with a short presentation of the concepts of the binary Boolean Algebra. Each reader should check his knowledge, particularly the knowledge of the different operation symbols and the applicable rules for the respective operations. Particularly the knowledge of the *antivalence* and the *equivalence* is very important, as it might not be as well known as the *conjunction*, the *disjunction*, and the *negation* (the *complement*).

Boolean equations and systems of Boolean equations are an important and elegant possibility to build models for very different problems. Here it is necessary to study the different solution methods. These solution methods are supported by the use of the XBOOLE system. It should be understood as an implementation and a summary of *Numerical Methods* for the binary Boolean Algebra. For larger numbers of variables where it is absolutely impossible to solve the problems "by hand," XBOOLE is a comfortable support for all necessary calculations. This includes also the operations of the Boolean Differential Calculus which will be introduced in the next chapter. It is necessary to be on the safe side when using this system.

1.2 BOOLEAN SPACE, BOOLEAN FUNCTIONS, AND OPERATIONS

In this chapter we repeat the basic concepts of the binary Boolean Algebra based on the set $B = \{0, 1\}$ with the two distinct elements 0 and 1. In technical applications 0 and 1 will be used all the time; in logics and in programming languages we also find **true** or **t** being used instead of 1 and **false** or **f** instead of 0. This will be indicated, if necessary, along with the respective applications.

We can define four functions from B into B:

x	$f_0(x)$	$f_1(x)$	$f_2(x)$	$f_3(x)$
0	0	0	1	1
1	0	1	0	1

The first function $f_0(x)$ is the constant function which is always equal to 0 (**contradiction**), the second function is the **identy** $f_1(x) = x$, the third function $f_2(x) = \overline{x}$ is called **negation**; it inverts the values into each other, i.e., 0 into 1 and 1 into 0. It will play an important role within the Boolean Differential Calculus. Finally, $f_3(x)$ is the function which is constantly equal to 1 (**tautology**). When we write the function values horizontally, i.e., (00), (01), (10), and (11), then we can understand the vectors as binary (dual) numbers, and the decimal equivalent of these numbers is equal to the index of the function.

With this set B, we now get

$$B^n = \underbrace{B \times \ldots \times B}_{n \text{ times}} = \{\mathbf{x} \,|\, \mathbf{x} = (x_1 \ldots x_n),\ x_i \in B,\ i = 1, \ldots, n\} \tag{1.1}$$

as the set of all binary vectors of length n (vectors with n components). It can easily be seen (by induction) that this set has 2^n elements. In Computer Science these vectors are well known as, for instance, *dual numbers* representing the natural numbers from 0 to $2^n - 1$.

Definition 1.1 Boolean function   Any unique mapping

$$f \,:\, B^n \Rightarrow B \tag{1.2}$$

is a *Boolean (logic) function* of n variables.

As a special case we use $n = 2$ and consider functions from B^2 into B which are called **conjunction** (*and*, $\wedge$), **disjunction** (*or*, $\vee$), **antivalence** (*exclusive-or*, $\oplus$), and **equivalence** (*if and only if*, $\odot$).

Table 1.1: The basic functions of two variables

x_1	x_2	$x_1 \wedge x_2$ conjunction	$x_1 \wedge x_2$ disjunction	$x_1 \oplus x_2$ antivalence	$x_1 \odot x_2$ equivalence
0	0	0	0	0	1
0	1	0	1	1	0
1	0	0	1	1	0
1	1	1	1	0	1

The set B^2 has four elements: (00), (01), (10), (11). When we consider these four vectors as the arguments of functions of two variables, then again the values 0 and 1 can be assigned to each vector, and therefore we have $2^4 = 16$ functions of two variables.

1.3 BOOLEAN FUNCTIONS AND OPERATIONS

Based on these definitions, logic functions can be described or defined by means of *expressions* (*formulas*). We start with variables $x_1, x_2, \ldots$. In a first step these variables can be negated, and we get $\overline{x}_1, \overline{x}_2$, etc. Variables and negated variables together are very often called *literals*. Each literal can take the values 0 or 1, resp. Now these literals can be combined by the given operations *conjunction, disjunction, antivalence,* and *equivalence,* resp., according to Tab. 1.1. In Tab. 1.1 we can see the following remarkable properties:

- The conjunction is equal to 1 if and only if both values are equal to 1.

- The disjunction is equal to 0 if and only if both values are equal to 0.

- The antivalence is equal to 0 if and only if both values are equal to each other.

- The equivalence is equal to 1 if and only if both values are equal to each other.

- The antivalence is equal to the negated equivalence (and vice versa).

  Other important rules are

- Negation rules:
$$x_1 \vee \overline{x}_1 = 1 , \quad x_1 \wedge \overline{x}_1 = 0 , \quad \overline{x}_1 = x_1 \oplus 1 = x_1 \odot 0 .$$

- De Morgan's Laws:
$$\overline{x_1 \wedge x_2} = \overline{x}_1 \vee \overline{x}_2 , \quad \overline{x_1 \vee x_2} = \overline{x}_1 \wedge \overline{x}_2 .$$

- Commutativity:
$$x_1 \vee x_2 = x_2 \vee x_1 , \quad x_1 \wedge x_2 = x_2 \wedge x_1 ,$$
$$x_1 \oplus x_2 = x_2 \oplus x_1 , \quad x_1 \odot x_2 = x_2 \odot x_1 .$$

- Associativity:
$$(x_1 \vee x_2) \vee x_3 = x_1 \vee (x_2 \vee x_3) , \quad (x_1 \wedge x_2) \wedge x_3 = x_1 \wedge (x_2 \wedge x_3) ,$$
$$(x_1 \oplus x_2) \oplus x_3 = x_1 \oplus (x_2 \oplus x_3) , \quad (x_1 \odot x_2) \odot x_3 = x_1 \odot (x_2 \odot x_3) .$$

- Distributivity:
$$x_1 \wedge (x_2 \vee x_3) = (x_1 \wedge x_2) \vee (x_1 \wedge x_3) , \quad x_1 \vee (x_2 \wedge x_3) = (x_1 \vee x_2) \wedge (x_1 \vee x_3) ,$$
$$x_1 \wedge (x_2 \oplus x_3) = (x_1 \wedge x_2) \oplus (x_1 \wedge x_3) , \quad x_1 \vee (x_2 \odot x_3) = (x_1 \vee x_2) \odot (x_1 \vee x_3) .$$

- Absorption:
$$x_1 \wedge (x_1 \vee x_2) = x_1 , \quad x_1 \vee (x_1 \wedge x_2) = x_1 .$$

- The use of 0 and 1:
$$x_1 \wedge 1 = x_1 , \quad x_1 \vee 1 = 1 , \quad x_1 \oplus 1 = \overline{x}_1 , \quad x_1 \odot 1 = x_1 ,$$
$$x_1 \wedge 0 = 0 , \quad x_1 \vee 0 = x_1 , \quad x_1 \oplus 0 = x_1 , \quad x_1 \odot 0 = \overline{x}_1 ,$$
$$x_1 \wedge \overline{x}_1 = 0 , \quad x_1 \vee \overline{x}_1 = 1 , \quad x_1 \oplus \overline{x}_1 = 1 , \quad x_1 \odot \overline{x}_1 = 0 .$$

- The elimination of $\oplus$ and $\odot$:
$$x_1 \oplus x_2 = (x_1 \wedge \overline{x}_2) \vee (\overline{x}_1 \wedge x_2), \quad x_1 \odot x_2 = (x_1 \wedge x_2) \vee (\overline{x}_1 \wedge \overline{x}_2).$$

Mostly the $\wedge$ between the variables or the brackets will be omitted (like the multiplication sign in arithmetic expressions). Therefore, $x_1 x_2$ means $x_1 \wedge x_2$, etc.

If necessary we can assume that two functions f and g always depend on the same set of variables. If, for instance, x_n is missing in the definition of a function f, then we duplicate the original definition of f for $x_n = 0$ and $x_n = 1$ using

$$f(x_1, \ldots, x_{n-1}) = f(x_1, \ldots, x_{n-1}) \wedge 1 = f(x_1, \ldots, x_{n-1})(x_n \vee \overline{x}_n), \qquad (1.3)$$

and the distributivity

$$\begin{aligned} f(x_1, \ldots, x_n) &= f(x_1, \ldots, x_{n-1})(x_n \vee \overline{x}_n) \\ &= x_n f(x_1, \ldots, x_{n-1}) \vee \overline{x}_n f(x_1, \ldots, x_{n-1}) . \end{aligned} \qquad (1.4)$$

In this way missing variables can be added for f and g until both functions depend on the same set of variables.

For the Boolean Differential Calculus operations will be used that are in a given sense reverse to this approach. We introduce the concepts of *subfunctions* and *subspaces*. The following example will show the approach quite easily. The given function depends on four variables, therefore, it is defined over B^4:

$$f(x_1, x_2, x_3, x_4) = x_1 \overline{x}_2 \overline{x}_3 x_4 \vee x_1 x_2 \overline{x}_3 \overline{x}_4 \vee x_1 \overline{x}_2 \overline{x}_3 \overline{x}_4 \vee \overline{x}_1 \overline{x}_2 \overline{x}_3 \overline{x}_4 . \qquad (1.5)$$

Now we set $x_1 = 0$ and get

$$f(0, x_2, x_3, x_4) = \overline{x}_2 \overline{x}_3 \overline{x}_4 \qquad (1.6)$$

as a **subfunction** of the function $f(x_1, x_2, x_3, x_4)$. It can also be said that this subfunction is considered in the subspace $(0, x_2, x_3, x_4)$ and it is therefore called *negative cofactor*. We receive a second subfunction by setting $x_1 = 1$:

$$f(1, x_2, x_3, x_4) = \overline{x}_2 \overline{x}_3 x_4 \vee x_2 \overline{x}_3 \overline{x}_4 \vee \overline{x}_2 \overline{x}_3 \overline{x}_4 . \qquad (1.7)$$

Now the subspace is equal to $(1, x_2, x_3, x_4)$ and $f(1, x_2, x_3, x_4)$ is called *positive cofactor*. It is quite easy to extend the procedure to more than one variable and build functions such as $f(1, x_2, 0, x_4)$, or $f(0, x_2, x_3, 1)$, or even $f(0, x_2, 0, 1)$, etc.

Up to this point in time, we used formulas in a rather intuitive way: the function was given by the function table, and the symbols like $\wedge$, $\vee$, etc. have been used as names of the functions defined in Tab. 1.1 to be used for the calculation of the function values. Mostly we used the *infix* notation for the functions, the symbol of the function has been written between the variables. Only for the complement the *prefix* notation has been used.

However, very often another approach is used. The formulas are the starting point, and the functions are introduced based on (meaningful) formulas. A formula will be considered, in this sense, as a set or sequence of computational rules by means of which one, some, or all functional values can be calculated. Many problems can be found because the meaning of a formula is not very clear: it cannot be seen whether the formula is a constant (equal to $0(\mathbf{x})$ or $1(\mathbf{x})$), how long it will take to calculate all the values, whether two different formulas express the same function, etc. In order to solve these problems appropriately, we need a precise definition of formulas that can be given in an inductive way.

Definition 1.2

1. $x_1, \ldots, x_n$, are formulas.

2. If F is a formula, then also $\overline{F}$.

3. If F_1, F_2 are formulas, then also

$$(F_1 \wedge F_2), (F_1 \vee F_2), (F_1 \oplus F_2), (F_1 \odot F_2) .$$

4. Any formula F can be constructed by a finite sequence of steps 1, 2, and 3 starting with the given variables $x_1, \ldots, x_n$.

Note. Rules for omitting brackets and operation symbols can be included and used as before. Only those expressions are *correct* formulas that can be built in this way. Sometimes it is necessary that each formula is a sequence of characters. Then the complement $\overline{F}$ has to be replaced by an appropriate symbol like $\neg F$ (or similarly).

The definition of a function can be built upon formulas using the *interpretation* of a formula. We used this concept already before; it consists in assigning values 0 or 1 to the variables and using the definitions of the operations (i.e., of the functions) that appear in the formula. If this has been done for each $\mathbf{x} \in B^n$, then the function has been computed and represented as a vector or a table.

Example 1.3   We explore the formula: $F = \overline{x}_1 \vee (x_1 \oplus x_2)(x_2 \oplus x_3)$. It can be seen that this formula can be built inductively:

1. x_1, x_2, x_3 are formulas.

2. $(x_1 \oplus x_2), (x_2 \oplus x_3)$ are formulas.

3. $((x_1 \oplus x_2)(x_2 \oplus x_3))$, $\overline{x}_1$ are formulas. The outer brackets of the first formula are omitted, $(x_1 \oplus x_2)(x_2 \oplus x_3)$ is a formula.

4. $(\overline{x}_1 \vee (x_1 \oplus x_2)(x_2 \oplus x_3))$ is a formula.

5. $\overline{x}_1 \vee (x_1 \oplus x_2)(x_2 \oplus x_3)$ is a formula; outer brackets can be omitted.

Interpretation: Let $\mathbf{x} = (101)$ be an assignment of Boolean values, then the following computations have to be performed:

$$\overline{\overline{1} \vee (1 \oplus 0)(0 \oplus 1)} \rightarrow \overline{\overline{1} \vee 1 \wedge 1} \rightarrow \overline{0 \vee 1} \rightarrow \overline{1} \rightarrow 0 .$$

The interpretation of this expression (formula) for each $\mathbf{x} \in B^3$ results in the following function:

x_1	x_2	x_3	$f(x)$
0	0	0	0
0	0	1	0
0	1	0	0
0	1	1	0
1	0	0	1
1	0	1	0
1	1	0	1
1	1	1	1

Sometimes it might be useful to include the constants 0 and 1 into the definition of formulas, by stating that $0, 1, x_1, x_2, \ldots, x_n$ are (initial) formulas. It can also be required that only some of the functions (sometimes only one single function) are to be used, etc.

We have already seen that different formulas can describe the same function. In fact, all the identities stated above give different formulas for the same function.

Definition 1.4   Two formulas F_1 and F_2 are equivalent if $f(F_1) = f(F_2)$.

$f(F_1)$ means the function f represented by the formula F_1. This definition implies that the same set of variables has to be considered for the two formulas (or functions). It can be shown that this relation between formulas is an equivalence relation, and, hence, the set of all formulas using a given fixed set of variables will be partitioned into equivalence classes. Every class contains the set of all expressions that represent the same function. Again, the use of dummy variables requires some attention, as above.

A formula F that represents the function $1(\mathbf{x})$ for a given $\mathbf{x}$ is called a *tautology* or a *logical law*.

Another comfortable way to represent a logic function is the enumeration of all vectors with $f = 1$ (or with $f = 0$) using a list. All the vectors that are not in the list will get the other value. We use, for instance, $n = 4$. Then we have, in principle, 16 binary vectors with four components. Now we can consider the following arrangement:

x_1	x_2	x_3	x_4	$f(x)$
1	0	0	1	1
1	1	0	0	1
1	0	0	0	1
0	0	0	0	1

Together with the setting $f = 1$ for these four vectors everything is done, the function is well defined. Now the following assignment is used:

$$1 \rightarrow x_i \qquad 0 \rightarrow \overline{x}_i . \tag{1.8}$$

By combining these settings by $\wedge$, four conjunctions are built and combined by $\vee$:

$$F(x_1, x_2, x_3, x_4) = x_1\overline{x}_2\overline{x}_3x_4 \vee x_1x_2\overline{x}_3\overline{x}_4 \vee x_1\overline{x}_2\overline{x}_3\overline{x}_4 \vee \overline{x}_1\overline{x}_2\overline{x}_3\overline{x}_4 . \tag{1.9}$$

This formula can be used as the definition of a function $f(x_1, x_2, x_3, x_4)$ which takes the value 1 exactly for the original four binary vectors. It is called *disjunctive normal form*.

The last two vectors have the common values $x_2 = 0, x_3 = 0, x_4 = 0$. They only differ in the first component: $x_1 = 0$ and $x_1 = 1$. Here we use the conjunction

$$\overline{x}_2\overline{x}_3\overline{x}_4 \qquad \text{and the \textbf{ternary vector}} \qquad (-000) .$$

The character '$-$' can be replaced by 0 or by 1. This means that such a ternary vector represents a *set of binary vectors* as well as a shorter conjunction. It reflects the following calculation (simplification):

$$x_1\overline{x}_2\overline{x}_3\overline{x}_4 \vee \overline{x}_1\overline{x}_2\overline{x}_3\overline{x}_4 = (x_1 \vee \overline{x}_1) \wedge \overline{x}_2\overline{x}_3\overline{x}_4 = 1 \wedge \overline{x}_2\overline{x}_3\overline{x}_4 = \overline{x}_2\overline{x}_3\overline{x}_4 .$$

We can also have a vector like $(-01-)$ which represents the conjunction $\overline{x}_2x_3$ as well as four binary vectors, etc.

For two ternary vectors s and t we can define the concept of **orthogonality**.

Definition 1.5 Orthogonality Two ternary vectors s and t are orthogonal to each other if there is at least one component i such that either $s_i = 0$ and $t_i = 1$ or $s_i = 1$ and $t_i = 0$.

This property has the enormous advantage that the set of binary vectors represented by the first vector and the set of binary vectors represented by the second vector have an *empty intersection*. These vectors will be the main data structure in the software package XBOOLE.

The listing of the vectors for $f = 0$ follows the same ideas.

1.4   BOOLEAN EQUATIONS

For two given functions $f(\mathbf{x})$ and $g(\mathbf{x})$ we consider the expression $f(\mathbf{x}) = g(\mathbf{x})$ as a Boolean equation, and each vector $\mathbf{x}$ with $f(\mathbf{x}) = g(\mathbf{x}) = 0$ or with $f(\mathbf{x}) = g(\mathbf{x}) = 1$ is a solution of the given equation. For an easy access to these concepts we will go back to the basic understanding of an equation, as it is known from elementary mathematics. An equation $ax + b = c$ or $ax^2 + bx + c = d$ is a constraint for the values of x, and a solution x_1 allows the transformation of the equations in the identities $c = c$ by calculating $ax_1 + b$ which results in c or by calculating $ax_1^2 + bx_1 + c$ that has to result in d. It is well known that these equations are equivalent to the equations $ax + (b - c) = 0$ or $ax^2 + bx + (c - d) = 0$ (i.e., these equations have the same solutions as the original equations). Further transformations result in $x + \frac{b-c}{a} = 0$ or $x^2 + \frac{b}{a}x + \frac{c-d}{a} = 0$ which can be changed to $x + x_0 = 0$ or $x^2 + px + q = 0$, and the solutions are

$$x = -x_0 , \qquad x_{1,2} = -\frac{p}{2} \pm \sqrt{\frac{p^2}{4} - q} ,$$

under consideration of several conditions, like $a \neq 0$, $\frac{p^2}{4} - q \geq 0$, etc.

In order to transform these ideas into the area of Boolean equations, we start with a problem that seems to look a bit strange. Let the equation

$$x_1 \vee x_2 = a \wedge b . \tag{1.10}$$

be given. Since we are dealing with logic functions, the identities can only have the format $0 = 0$ or $1 = 1$. $x_1 \vee x_2 = 0$ holds only for $x_1 = 0, x_2 = 0$. $a \wedge b = 0$ holds for $a = 0$, $b = 0$; $a = 0$, $b = 1$; $a = 1$, $b = 0$; hence, we have the following set of solution vectors with the components (x_1, x_2, a, b):

$$\{(0000), (0001), (0010)\} .$$

Now the identity $1 = 1$ has to be explored, and, according to the definition of $\vee$ and $\wedge$, we get the following solutions:

$$\{(0111), (1011), (1111)\} .$$

Altogether, this equation has six solution vectors. The strange character of this equation comes from the fact that there are different variables on the left side and the right side of the equation. In order to equalize the variables on both sides, we use dummy variables:

$$(x_1 \vee x_2)(a \vee \overline{a})(b \vee \overline{b}) = a b (x_1 \vee \overline{x}_1)(x_2 \vee \overline{x}_2) ,$$

and this finally results in the equation

$$x_1\,a\,b \vee x_2\,a\,b \vee x_1\,a\,\overline{b} \vee x_2\,a\,\overline{b} \vee x_1\,\overline{a}\,b \vee x_2\,\overline{a}\,b \vee x_1\,\overline{a}\,\overline{b} \vee x_2\,\overline{a}\,\overline{b} =$$
$$a\,b\,x_1\,x_2 \vee a\,b\,x_1\,\overline{x}_2 \vee a\,b\,\overline{x}_1\,x_2 \vee a\,b\,\overline{x}_1\,\overline{x}_2 .$$

By checking all 16 vectors of B^4, it can be seen that this equation has the same solutions as the original equation $x_1 \vee x_2 = a\,b$. Hence, we can assume that in an equation $f(\mathbf{x}) = g(\mathbf{x})$ the functions f and g depend on the same set of variables.

Definition 1.6   Boolean equation   Let $\mathbf{x} = (x_1, \ldots, x_n)$, $f(\mathbf{x})$, $g(\mathbf{x})$ be two Boolean (logic) functions, then

$$f(\mathbf{x}) = g(\mathbf{x}) \tag{1.11}$$

is a *Boolean (logic) equation* of n variables. The vector $\mathbf{b} = (b_1, \ldots, b_n)$ is a *solution* of this equation if $f(\mathbf{b}) = g(\mathbf{b})$ (i.e., $f(\mathbf{b}) = g(\mathbf{b}) = 0$ or $f(\mathbf{b}) = g(\mathbf{b}) = 1$).

In the same way as above, it is possible to reduce the considerations to homogeneous equations.

Theorem 1.7   Homogeneous equations   *The equation $f(\mathbf{x}) = g(\mathbf{x})$ is equivalent to Eqs. (1.12) and (1.13).*

$$f(\mathbf{x}) \oplus g(\mathbf{x}) = 0 \tag{1.12}$$
$$f(\mathbf{x}) \odot g(\mathbf{x}) = 1 \tag{1.13}$$

Proof. This theorem follows directly from the definition of $\oplus$ and $\odot$.   □

1.5   SYSTEMS OF EQUATIONS AND INEQUALITIES

In contrast to other areas of mathematics, systems of Boolean (logic) equations can be reduced to one single equation.

Theorem 1.8   Systems of Boolean equations   *Let*

$$f_1(\mathbf{x}) = g_1(\mathbf{x})\,,$$
$$\vdots$$
$$f_m(\mathbf{x}) = g_m(\mathbf{x}) \tag{1.14}$$

be a system of m equations of n variables. Then Eq. (1.14) is equivalent to Eqs. (1.15) and (1.16).

$$\{f_1(\mathbf{x}) \oplus g_1(\mathbf{x})\} \vee \ldots \vee \{f_m(\mathbf{x}) \oplus g_m(\mathbf{x})\} = 0 \tag{1.15}$$
$$\{f_1(\mathbf{x}) \odot g_1(\mathbf{x})\} \wedge \ldots \wedge \{f_m(\mathbf{x}) \odot g_m(\mathbf{x})\} = 1 \tag{1.16}$$

Proof. This theorem follows directly from the definition of $\oplus$, $\odot$, $\wedge$, $\vee$.   □

In a first step, all the equations are transformed into the homogeneous form ($= 0$ or $= 1$); these homogeneous equations are then transformed into one equation according to

$$a = 0, \quad b = 0 \quad \Leftrightarrow \quad a \vee b = 0\,,$$
$$a = 1, \quad b = 1 \quad \Leftrightarrow \quad a \wedge b = 1\,.$$

This means that, at least in principle, only single homogeneous equations of n variables have to be considered. The practical way will very often have the reverse direction: a very complicated equation will be split into "smaller" parts, the solution sets of these parts will be found and thereafter combined to the solution set of the given equation. This approach will be explained in detail in the next section.

It is also not very difficult to deal with *inequalities*. Two cases have to be considered (according to the definition of $\leq$ or $<$ in B):

$$f(\mathbf{x}) \leq g(\mathbf{x})\,, \tag{1.17}$$
$$f(\mathbf{x}) < g(\mathbf{x})\,. \tag{1.18}$$

Ineq. (1.17) has the solutions $f = 0$, $g = 0$; $f = 0$, $g = 1$; $f = 1$, $g = 1$. This solution set can be translated into one equation with the same solution set (i.e., into an equivalent equation):

$$\overline{f(\mathbf{x})} \vee g(\mathbf{x}) = 1 \Leftrightarrow f(\mathbf{x})\,\overline{g(\mathbf{x})} = 0\,. \tag{1.19}$$

Ineq. (1.18) has only one solution $f = 0$, $g = 1$ which can be expressed by the equation(s)

$$f(\mathbf{x}) \vee \overline{g(\mathbf{x})} = 0 \Leftrightarrow \overline{f(\mathbf{x})}g(\mathbf{x}) = 1\,. \tag{1.20}$$

This is already a proof of Th. 1.9.

Theorem 1.9   *Each inequality $f(\mathbf{x}) \leq g(\mathbf{x})$ and $f(\mathbf{x}) < g(\mathbf{x})$ can be transformed into equivalent homogeneous equations:*

$$f(\mathbf{x}) \leq g(\mathbf{x}) \Leftrightarrow \overline{f(\mathbf{x})} \vee g(\mathbf{x}) = 1 \Leftrightarrow f(\mathbf{x})\overline{g(\mathbf{x})} = 0\,, \tag{1.21}$$
$$f(\mathbf{x}) < g(\mathbf{x}) \Leftrightarrow f(\mathbf{x}) \vee \overline{g(\mathbf{x})} = 0 \Leftrightarrow \overline{f(\mathbf{x})}g(\mathbf{x}) = 1\,. \tag{1.22}$$

Hence, homogeneous equations also cover inequalities between Boolean functions as well as systems of inequalities and all kinds of *mixed* systems.

1.6   SOLUTIONS WITH REGARD TO VARIABLES

A last very important concept, possibly the most important altogether in this area, is the solution of logic equations **with regard to variables**. As a motivation, we go back to the equation $x^2 + px + q = 0$. Previously we had the understanding that p and q are constant real numbers, and for each pair of these numbers, we can explore whether the condition $\frac{p^2}{4} - q \geq 0$ is satisfied, and if so, the solutions x_1 and x_2 can be calculated. However, we also can use another point of view. We think about $x^2 + px + q$ as a function $f(x, p, q)$ of three variables and ask whether the equation $x^2 + px + q = 0$ defines, for instance, a function $x(p, q)$ that satisfies the equation, or whether functions $p(x)$ and $q(x)$ can be found satisfying this equation, etc. In order to do this the equation $x^2 + px + q = 0$ can be used to find all the points (x, p, q) in a three-dimensional space satisfying the given equation. Thereafter, we can try to find such functions $x(p, q)$ or pairs $p(x), q(x)$ satisfying the equation, etc.

This problem exists also for Boolean equations and has many applications. As we have already seen, one equation can be considered as the *Summary* of many equations, inequalities, etc. All the knowledge existing for a given binary problem can be put together, and then we can try to find such implicitly given functions that describe the most interesting parts of the *logic model*. The applications will show the working of this methodology. Here we introduce the required concepts for the Boolean equations.

At the beginning let the equation $f(x_1, \ldots, x_n) = 0$ be given and the solution set S. The set of variables $\{x_1, \ldots, x_n\}$ can be partitioned into two disjoint subsets $\{x_1, \ldots, x_k\}$ and $\{x_{k+1}, \ldots, x_n\}$. This partition also partitions each binary vector $\mathbf{b}$ of the solution set into two parts $(b_1, \ldots, b_k)$ and $(b_{k+1}, \ldots, b_n)$. The equation defines a mapping φ from B^k into B^{n-k}. Each vector $\mathbf{x}$ has been assigned to one, more than one, or even zero vectors of B^{n-k}, and this happens for any of the partitions of the set $\{x_1, \ldots, x_n\}$. In order to easily understand the role of different variables, we write $f(x_1, \ldots, x_k, y_1, \ldots, y_m)$, but we keep in mind that any partition of $\{x_1, \ldots, x_n\}$ can be considered. Very often the role of the different variables will be defined by the context of the problem. It can also be said that the equations define a given relation between the vectors of B^k and B^m.

Especially interesting are now the cases where the mapping φ is unique and where φ is complete. In the first case every $\mathbf{x} \in B^k$ has been assigned at most one element of B^m, however, some $\mathbf{x}$ might not have an assigned element. In the second case each $\mathbf{x} \in B^k$ will have an assigned element of B^m, however, since this is not necessarily defined in a unique way, more than one element is quite possible. These two cases will be considered for the solution with regard to variables.

Definition 1.10 Let $\mathbf{x} = (x_1, \ldots, x_k), \mathbf{y} = (y_1, \ldots, y_m), f(\mathbf{x}, \mathbf{y})$ be a function of $n = k + m$ variables. The equation $f(\mathbf{x}, \mathbf{y}) = 0$ can be solved with regard to the variables (functions) $y_1, \ldots, y_m$ if there are functions

$$y_1 = g_1(\mathbf{x}),$$
$$\vdots$$
$$y_m = g_m(\mathbf{x}), \tag{1.23}$$

with

$$f(\mathbf{x}, g_1(\mathbf{x}), \ldots, g_m(\mathbf{x})) = 0. \tag{1.24}$$

Table 1.2: The solution with regard to variables

x_1	x_2	y_1	y_2	$f(x_1,x_2,y_1,y_2)$	1	2	3	4	5	6	7	8
0	0	0	0	0	00	00	00	00				
0	0	0	1	0					01	01	01	01
0	1	0	0	0	00	00			00	00		
0	1	0	1	0			01	01			01	01
1	0	1	0	0	10		10		10		10	
1	0	1	1	0		11		11		11		11
1	1	1	1	0	11	11	11	11	11	11	11	11

Example 1.11

1. *Completely but not uniquely solvable equation with regard to* y_1, y_2:

The equation

$$f = \overline{x}_1\, y_1 \vee x_1\, \overline{y}_1 \vee x_2\, y_1\, \overline{y}_2 = 0 \tag{1.25}$$

has to be solved with regard to the variables y_1 and y_2. Tab. 1.2 shows the solutions of this equation and the different pairs of functions $(y_1(x_1, x_2), y_2(x_1, x_2))$ that solve this equation.

The mapping from B^2 into B^2 is complete, the domain of the mapping is the whole B^2, but it is not unique, three of the four **x**-vectors have two elements assigned to them. The grouping of the solutions follows the values of the **x**-vectors. Hence, in order to define the functions $y_1(\mathbf{x})$, $y_2(\mathbf{x})$, for $\mathbf{x} = (00)$ two vectors for (y_1, y_2) can be selected, the same applies to $\mathbf{x} = (01)$ and $\mathbf{x} = (10)$; only the value for $\mathbf{x} = (11)$ is defined in a unique way. The right side of Tab. 1.2 shows the selections.

The eight different columns define eight different pairs of functions: $y_1 = x_1$ is the only possibility for y_1 (defined by the first component of the pairs), for y_2 we find

$$
\begin{array}{ll}
1: \ y_2 = x_1\, x_2\,, & 5: \ y_2 = x_1 \odot x_2\,, \\
2: \ y_2 = x_1\,, & 6: \ y_2 = x_1 \vee \overline{x}_2\,, \\
3: \ y_2 = x_2\,, & 7: \ y_2 = \overline{x}_1 \vee x_2\,, \\
4: \ y_2 = x_1 \vee x_2\,, & 8: \ y_2 = 1\,,
\end{array}
$$

defined by the second component in the eight columns. Let us use, for instance, $y_1 = x_1$, $y_2 = x_1 \vee \overline{x}_2$, then we get for (1.25):

$$f = \overline{x}_1\, y_1 \vee x_1\, \overline{y}_1 \vee x_2\, y_1\, \overline{y}_2 = \overline{x}_1\, x_1 \vee x_1\, \overline{x}_1 \vee x_2\, x_1\, \overline{x}_1\, x_2 = 0 \vee 0 \vee 0 = 0\,.$$

2. *Completely and uniquely solvable equation with regard to y_1, y_2:*

The equation

$$f = y_1 (x_1 x_2 \vee \overline{x}_1 \overline{x}_2) \vee \overline{y}_1 (x_1 \overline{x}_2 \vee \overline{x}_1 x_2) \vee y_2 \overline{(x_1 \vee x_2)} \vee \overline{y}_2 (x_1 \vee x_2) = 0 \qquad (1.26)$$

defines the structure shown in Tab. 1.3. The mapping φ is complete and also unique. The pair $y_1 = x_1 \oplus x_2$ and $y_2 = x_1 \vee x_2$ is the only pair of solution functions.

The substitution of these functions into (1.26) leads to the identity:

$$\begin{aligned}
f &= y_1 (x_1 x_2 \vee \overline{x}_1 \overline{x}_2) \vee \overline{y}_1 (x_1 \overline{x}_2 \vee \overline{x}_1 x_2) \vee \\
&\quad y_2 \overline{(x_1 \vee x_2)} \vee \overline{y}_2 (x_1 \vee x_2) \\
&= (x_1 \oplus x_2)(x_1 x_2 \vee \overline{x}_1 \overline{x}_2) \vee \overline{(x_1 \oplus x_2)}(x_1 \overline{x}_2 \vee \overline{x}_1 x_2) \vee \\
&\quad (x_1 \vee x_2) \overline{(x_1 \vee x_2)} \vee \overline{(x_1 \vee x_2)}(x_1 \vee x_2) \\
&= (x_1 \oplus x_2) \overline{(x_1 \oplus x_2)} \vee \overline{(x_1 \oplus x_2)}(x_1 \oplus x_2) \vee \\
&\quad (x_1 \vee x_2) \overline{(x_1 \vee x_2)} \vee \overline{(x_1 \vee x_2)}(x_1 \vee x_2) \\
&= 0 \vee 0 \vee 0 \vee 0 \\
&= 0 .
\end{aligned}$$

Table 1.3: The unique solution with regard to variables

x_1	x_2	y_1	y_2	$f(x_1, x_2, y_1, y_2)$	1
0	0	0	0	0	00
0	1	1	1	0	11
1	0	1	1	0	11
1	1	0	1	0	01

3. *Uniquely but not completely solvable equation with regard to y_1, y_2:*

Finally we consider the equation

$$f = x_1 \overline{y}_2 \vee \overline{x}_1 y_2 \vee x_1 x_2 \vee \overline{y}_1 y_2 \vee x_2 \overline{y}_1 \vee \overline{x}_1 \overline{x}_2 y_1 = 0 . \qquad (1.27)$$

This equation has only three solutions. There is no solution with $x_1 = 1$, $x_2 = 1$. The mapping φ is unique, however, it is not complete.

When we are using the functions $y_1 = x_1 \oplus x_2$ and $y_2 = x_1 \overline{x}_2$, we describe the correct solutions, but we must exclude $x_1 = 1$, $x_2 = 1$. Hence, we define the constraint $x_1 x_2 = 0$ that defines the three allowed **x**–vectors (00), (01), (10), and we can say that the given functions are solutions under the assumption that the constraint is satisfied as well. This kind of solution with regard to variables can always be achieved, except for the function $1(\mathbf{x})$.

Table 1.4: The solution with regard to variables under constraints

x_1	x_2	y_1	y_2	$f(x_1, x_2, y_1, y_2)$
0	0	0	0	0
0	1	1	0	0
1	0	1	1	0

In the given examples the role of the $x_1, x_2, \ldots$ and $y_1, y_2, \ldots$ has already been defined at the beginning. In many applications, however, we will solve an equation $f(\mathbf{x}) = 0$, and consequently, we try to find a minimal number of variables $x_{i_1}, \ldots, x_{i_k}$, so that the other remaining variables are *functionally dependent* on the *independent variables*. A small number of independent variables is highly desirable in many cases. Especially for diagnostic situations this approach is very useful. The following theorem is very useful to define the number of independent variables that can be achieved.

Theorem 1.12   *The mapping $\varphi : B^k \to B^m$ can be complete only if the equation $f(\mathbf{x}, \mathbf{y}) = 0$ has at least 2^k solutions. The mapping $\varphi : B^k \to B^m$ can be unique and complete only if the number of solutions of the equation $f(\mathbf{x}, \mathbf{y}) = 0$ is equal to 2^k.*

Proof. When the number of solutions of the equation $f(\mathbf{x}, \mathbf{y}) = 0$ is less than 2^k then not each $\mathbf{x}$ can appear in the solution set, hence, the mapping φ will not be complete. When the mapping $\varphi : B^k \to B^m$ is supposed to be unique and complete, then each vector $\mathbf{x}$ appears once and only once in the domain of φ, hence, the number of solutions must be equal to 2^k.   □

Another interesting possibility is the introduction of independent parameters $t_1, t_2, \ldots$ and the expression of all the solutions by means of these parameters.

Let $f = (x_1 \vee x_2)(y_1 \oplus \overline{y}_2) \vee \overline{x}_1 \overline{x}_2 (y_1 \oplus y_2)$ be given. The equation $f = 0$ has eight solutions $\{(0000), (0011), (0101), (0110), (1101), (1110), (1001), (1010)\}$ that can be *coded* by three parameters t_1, t_2, t_3.

$$x_1 = t_1 , \quad x_2 = t_1 \oplus t_2 , \quad y_1 = t_3 , \quad y_2 = (t_1 \vee t_2)\overline{t}_3 \vee \overline{t}_1 \overline{t}_2 t_3$$

define the vectors of the solution set by means of four functions depending on the independent parameters t_1, t_2, t_3.

The introduction of independent variables is very elegant and efficient. In many applications the equation (the constraint) is known, and after the solution of the equation, the solution set can be described (stored) by a number of functions depending directly on a smaller number of parameters. Even the reproduction of the solution set by means of circuits or programs is quite feasible and can use this approach.

1.7   XBOOLE

1.7.1   CONCEPT AND PROPERTIES

In the 1980s we developed the XBOOLE system, in the sense of *numerical methods* for logic functions, because in practical applications the number of variables the logic functions are depending on is very large. Therefore, hundreds or thousands of logic values must be manipulated without any error. Obviously, computers and suitable software are required to solve realistic logic problems.

A Boolean variable can carry only one of the two values 0 or 1 at one point in time. Hence, Boolean variables are the simplest variables of all. Programming languages provide Boolean operations for Boolean variables and limited vectors of Boolean values. These vectors can be used to store the function values of a Boolean function in a fixed order. However, the number of function values of a Boolean function of n variables is equal to 2^n. Due to this exponential complexity the direct calculation of Boolean functions with the elements of a programming language is strongly limited.

The aim of XBOOLE is to bridge the gap between the easy and efficient calculation of large Boolean functions and the restricted basis of the programming languages. XBOOLE is not focused on solving a special problem. Sets of binary vectors as a model of many different applications are the theoretical basis.

Two binary vectors which differ in a single component can be merged into one ternary vector that contains a dash-element $(-)$ instead of the 0 or 1. A single ternary vector with d dashes represents 2^d binary vectors. These ternary vectors are the main data structure of XBOOLE. The exponential increase of the number of binary vectors with the linear increase of the number of Boolean variables is curbed by the merging of a set of binary vectors into a single ternary vector. All 2^n binary vectors of n variables are expressed by a single ternary vector of n dashes.

Several ternary vectors which belong semantically together are combined in XBOOLE into a *ternary matrix* (TM) which is also called *ternary vector list* (TVL). All ternary vectors of a TVL are orthogonal to each other when each binary vector belongs at most to one ternary vector of this TVL. Such a TVL is called *orthogonal* TVL. XBOOLE prefers orthogonal TVL but can handle also TVLs which are not orthogonal.

Generally, each TVL depends on different sets of Boolean variables in any order. Calculations with such TVLs require the assignment of associated columns of the same variable. This is a very time-consuming procedure. XBOOLE solves this problem by the so-called *space concept*. This concept allows us to define an arbitrary number of Boolean spaces with any number of Boolean variables which can be chosen for each space. In this way, the number of Boolean variables is not limited. Within each Boolean space the variables can be explicitly assigned in a wanted order implicitly assigned by the XBOOLE system. XBOOLE operates with TVL which belong to the same space. Hence, time-consuming adaptions of columns are avoided. One XBOOLE operation allows the transfer of any TM between two different Boolean spaces.

One more source of the efficiency of XBOOLE is that not the values of single Boolean variables but all Boolean values of a whole machine word are calculated in parallel. A sequence of such parallel calculation steps is necessary when the number of Boolean variables is larger than the word length of the computer. In many cases XBOOLE detects that this sequential procedure can be truncated such that the parallel calculations generally dominate.

The required memory to store a TVL varies in a wide range depending on the number of variables of the chosen Boolean space and the number of ternary vectors. XBOOLE solves this problem by the so-called *box concept*. XBOOLE divides an adequately large part of the main memory into boxes of a fixed size and uses such boxes as many as necessary to store a TVL. All calculations are realized by XBOOLE directly in these boxes such that no load or store operations are needed as pre- or post-processing steps.

These main concepts provide a wide range of logic operations and are realized in about hundred functions written in the programming language C. Hence, the XBOOLE Library can be used as a valuable extension of C of C++ programs in the Boolean domain. Sec. 1.7.4 shows this kind of application.

In order to offer many users of different application domains the access to XBOOLE, we developed a tool which is called *XBOOLE Monitor*. The XBOOLE Monitor can be considered as a *logic pocket calculator*. Almost all functions of the XBOOLE Library are wrapped by the XBOOLE Monitor such that they can be easily used and combined to solve logic problems. Sec. 1.7.2 shows how the XBOOLE Monitor can be downloaded for free and how the appropriate operations can be executed in order to get the required result. We will use the XBOOLE Monitor in this book to solve Boolean differential equations.

1.7.2   XBOOLE MONITOR

We use in this book the XBOOLE Monitor xbm32.exe that provides a graphical user interface and runs under several versions of the Windows Operating System. This XBOOLE Monitor can be downloaded by everybody for free from the following web page:

http://www.informatik.tu-freiberg.de/xboole.

At the left side of this page the link XBOOLE Monitor can be seen. A click on this link leads to the download page of the XBOOLE Monitor. In order to download the XBOOLE Monitor, the button with the label XBOOLE Monitor located at the bottom of this page must be pressed to start the standard download procedure. After the download, the file XBOOLEMonitor.zip must be unzipped into the new directory XBOOLEMonitor. Tab. 1.5 shows the unzipped files of the XBOOLE Monitor.

Tab. 1.5 shows that the XBOOLE monitor can be used in English or German. The executable file of the XBOOLE Monitor is xbm32.exe. It can be started without any further installation, and it automatically selects the language of all the representations and the help environment depending on the language of the Windows Operating System that is used on the respective machine.

Fig. 1.1 shows a screenshot of the XBOOLE Monitor the size of which has been strongly minimized in order to meet the size of the pages of this book. The size of this window can be resized in the same way as any other window on the screen.

Table 1.5: Meaning of the files in the directory `XBOOLEMonitor`

Name of the File	Meaning of the File
xbm32.cnt	content structure of the help file in use
xbm32.exe	executable file of the XBOOLE Monitor
xbm32.hlp	help file in use
xbm32_e.cnt	content structure of the English help file
xbm32_e.hlp	help file in English
xbm32_g.cnt	content structure of the German help file
xbm32_g.hlp	help file in German

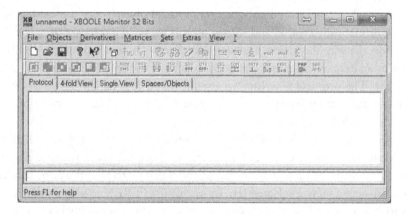

Figure 1.1: Complete window structure of the XBOOLE Monitor.

The window of the XBOOLE Monitor comprises several parts. In the following the main purpose of these parts will be explained.

The headline of the XBOOLE Monitor window shows the icon of this program, the name of the file opened or stored last, and the title of the program *XBOOLE Monitor 32 Bits*. The term unnamed as the name of the file means that no file has been used so far. Otherwise a file name with the extension sdt is shown. Such an sdt-file allows us to interrupt the work with the XBOOLE Monitor. After loading a stored sdt-file the work with the XBOOLE Monitor based on previous data can be continued.

The menu bar is located below the headline of the XBOOLE Monitor. The menu is structured like a tree and allows the complete control of the behavior of the XBOOLE Monitor. Typically, the menu items will be selected using the mouse. Alternatively, the ALT-key in connection with further keys of the keyboard can be used. Additional information for the selected action must be specified in most cases in special dialog windows.

A set of toolbars is located below the menu bar. Each toolbar can be shown or hidden separately. Its position can also be arranged by the user. All these adjustments are automatically stored in the Windows Registry database such that the adjustment of the toolbars does not change after the next start of the XBOOLE Monitor. The toolbars were introduced to shorten the management of the XBOOLE Monitor. Instead of selecting the whole path between the root and the leaf in the menu tree step by step, the action associated to the wanted leaf can be selected by a single click on the associated icon in a toolbar. Note: toolbar icons are defined only for actions which are frequently used. Hence, the actions controllable by the toolbars are a subset of all available actions.

The part below the toolbars covers the main space of the XBOOLE Monitor window. The XBOOLE Monitor visualizes different types of information in this area, but also allows partial interaction with the user. This part is structured by a tab control. A click with the mouse brings the page associated to the tab in the foreground.

The first tab is labeled by `Protocol`. A protocol of each action executed by the XBOOLE Monitor is automatically created and shown on this page. The representation of the protocol does not depend on the way the action has been initialized. There are three possibilities to activate an action: first by an item of the menu, second by an icon of the toolbar, and third by a command of the command line. The protocol is written as a sequence of commands. There is a possibility to store the protocol and execute this sequence of commands again at a later point in time. Such a sequence of commands is called *problem program* or shortly PRP.

The second tab with the label `4-fold View` and the third tab with the label `Single View` provide basically the same behavior. The difference between these pages is that the 4-fold view consists of a 2x2-matrix, where each separate sheet of this matrix has the same behavior as the whole single view. In such a view a selected ternary vector list (TVL), an associated Karnaugh map, or a selected variable tuple (VT) can be shown. There is an edit mode in such a view which allows us to edit the elements of a TVL. If there is not enough space, scroll bars appear automatically and allow the selection of each part of the represented data. It is suggested to use the single view for very large objects. Otherwise, the 4-fold view is preferred, because four objects can be seen at the same time.

Fig. 1.2 shows an example of the 4-fold view of the XBOOLE Monitor. The top left part shows the first object, a TVL in disjunctive (D) form, depending on three variables, consisting of three rows, and defined in the first space. The top right part shows the Karnaugh map of this Boolean function. The button labeled by K in the TVL-representation switches to the Karnaugh map, and the button labeled by T in the Karnaugh map representation to the TVL, respectively. The bottom left part shows the second object, a VT of two variables in the first space. The bottom right part shows the third object, a TVL in orthogonal disjunctive/antivalence (ODA) form, depending on three variables, consisting of two rows, and defined in the first space. This TVL was calculated as the complement of the first TVL.

The forth tab is labeled by `Spaces/Objects`. This page is divided vertically into two views. The left view shows a list of all details of the Boolean spaces defined by the user. The right view shows a list of all TVLs and VTs recently stored in the XBOOLE Monitor, where each of these

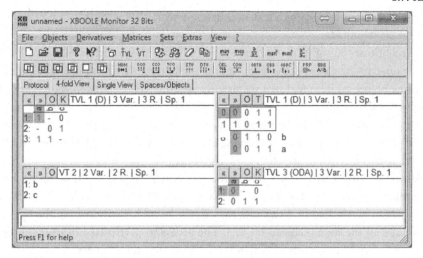

Figure 1.2: Example of the 4-fold view in the XBOOLE Monitor.

objects is uniquely identified by its object number. We refer in this book to the XBOOLE object with the number i by $XBO[i]$. In addition to the number of the object and its type information about the form, the space, and numbers of variables, rows, and boxes are given.

The command line is located below the main part of the XBOOLE window. While the menu bar and the toolbar allow simple intuitive operations of the XBOOLE Monitor, the application of the command line requires the knowledge of the commands for the control of the actions. The benefit of the command line is that XBOOLE operations can be activated faster, because interactions in dialog windows are omitted. The details of all commands can be studied in the XBOOLE help system.

The bottom of the XBOOLE window contains the status bar. In this bar help information about the action selected by a menu item or by a toolbar icon will be displayed.

The menu allows us to activate each action executable by the XBOOLE Monitor as usual. It has the structure of a tree. The items directly connected to the root are shown in the menu bars visible in Fig. 1.1 and Fig. 1.2.

Some of the actions require a precondition for their execution. It is, for instance, not possible to calculate the complement of a function if no function has been defined before. In such cases the menu item is deactivated by the XBOOLE Monitor in order to protect the XBOOLE user from errors. If a precondition does not hold, then the associated menu item appears in grey, and its activation is not possible.

Three points (...) behind the text of a menu item indicate that a dialog window appears which allows the input of some required information for the selected action. It is, for instance, possible to define the basic settings for the program by selecting the item Settings... in the submenu File.

A click on this menu item opens the dialog window which allows us to configure the appearance of the XBOOLE Monitor.

A triangle at the right of the menu item indicates that another submenu is associated. This exists only once in the menu View where a next menu opens while selecting the item Toolbars.

A tick before an item text in a menu indicates that the associated state is true. Such ticks are used in the View menu in order to indicate which parts are visible at present. For instance, one click onto the item Status Bar in the menu View removes the status bar and the tick associated to this item. The next click on the same item changes the state of the status bar again, and this means that the status bar and the tick in the item is again visible.

All the details of the application of the XBOOLE Monitor can be found in the integrated help environment. There are two methods in order to get help information. First, the complete help information can be studied in a separate dialog window. This window will be opened by the item Help Topics in the menu labeled by '?'. The access to particular information topics is possible using structured contents, a predefined index, or a supported search across the whole help text. The second help method works context-sensitively. After clicking the item Context Help in the menu labeled by '?', a special cursor that includes a question mark appears. A click with this cursor on a menu item opens a help window that includes the help information associated to the selected item.

In the initial state of the XBOOLE Monitor all items in the menus Derivative, Matrices, and Sets are deactivated, indicated by grey items, because so far no TVLs exist. A TVL can be created using the item Create TVL... in the menu Objects, but this item is deactivated in the initial state of the XBOOLE Monitor as well, because up to now no Boolean space has been defined. Hence, it is necessary that the user of the XBOOLE Monitor defines at least one Boolean space.

The XBOOLE Monitor allows calculations for Boolean functions of an unlimited number of variables. These variables must be associated with Boolean spaces where the number of variables in each Boolean space is restricted by the user of the XBOOLE Monitor who defined such spaces. In order to define a new Boolean space, the item Define Space... in the menu Objects can be used. Fig. 1.3 shows the dialog window that appears after selecting this item. In the upper edit control the number of the Boolean space to be created is suggested and can be changed into a new number which has not been used so far as a space number. In the lower edit control the maximum number of variables in the range of 1 to 1952 can be specified. Note: this value cannot change later on. The suggested number of variables (32) fits to the word length of the CPU and is generally a good choice.

The definition of a Boolean space defines only the number of variables, but not their names nor the order of their appearance. Very often, these names and their order are derived from the input, for instance, of an equation. However, sometimes it might be useful to display the variables in a given order, for instance, first the input variables and thereafter the output variables, or similarly. In order to achieve this, the names and the order of variables in a given space can be defined after the definition of the space, but before the input of a TVL. This can be achieved by using the menu items Attach Variables... or Append Variable(s) to a VT...

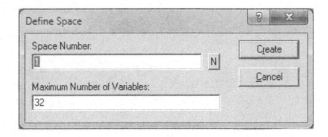

Figure 1.3: Dialog window that allows the definition of a new Boolean space.

The toolbars allow a direct activation of a subset of actions of the menu. This speeds up the activation of XBOOLE actions, but requires the knowledge of the toolbar icons. The behavior of the toolbars is in agreement with the menu. An icon of a toolbar can be activated by a left mouse click only if the precondition of the associated action is satisfied. Otherwise, the icon of a toolbar is deactivated, which is visible by a grey icon.

There are six toolbars. Using the item `Toolbars` in the menu `View`, it is possible to show or to hide each bar separately. The position of each toolbar can change by drag and drop. Both Fig. 1.1 and Fig. 1.2 show the toolbars comfortably arranged in two rows.

The first toolbar `General` covers selected actions of menus labeled by `File` and '`?`'. The toolbar `Objects` covers seven of ten actions of the menu `Objects`. All actions of the menus `Derivatives`, `Matrices`, and `Sets` are typically used very often. Therefore, all these actions can be activated directly by icons of the associated toolbars `Derivatives`, `Matrices`, and `Sets`, respectively. The last toolbar `Extras` covers two of six actions of the menu labeled by `Extras`.

The meaning of the toolbar icons can be studied easily. The simplest way is to move the mouse pointer over an icon of a toolbar. Immediately a small yellow window appears which shows a quick info text that explains the meaning of the icon. Additionally, a more detailed description is given in the status bar. Well-structured, complete information about the meaning of the icons in all toolbars can be obtained by using the help environment of the XBOOLE Monitor. In order to do this, the item `Help topics` can be used as described above, or simply the F1 button can be pressed. The item `toolbars` in the first case or the link `Summary of the toolbars` lead to the same information page `Survey of the Toolbars` where information about each toolbar can be selected.

As an example for the application of toolbars, TVL 1 of Fig. 1.2 will be created. In order to do this, it is assumed that space 1 has been created. In order to activate the action that creates a TVL, it is sufficient to click with the left mouse button on the second icon of the toolbar `Objects`. This icon is labeled by `+ TVL`. All the information about the TVL which will be presented now is transmitted to the XBOOLE Monitor in the same way as it is done for the initialization of a TVL using the item `Create TVL...` of the menu `Objects`.

The action to create a TVL first opens a dialog window shown in Fig. 1.4. Four properties of the TVL to be created must be defined in this dialog window. In a first step the Boolean space

Figure 1.4: Dialog window to specify basic properties of a new TVL.

which comprises the TVL must be selected from a list. The next step defines an object number for this new TVL; this number can be used in the future to call this TVL. Note: if an object (TVL or VT) with this number already exists, the old object will be deleted. A simple way to select a new number (not yet used) consists in pressing the button labeled by N in the dialog window. In a third step the form of the TVL must be selected from a list of possible forms. The selected form D means that the TVL will be given in disjunctive form. Note: if the form ODA is selected, the input TVL will be taken as a D-form and orthogonalized internally. In a last (fourth) step the variables of the TVL columns must be written into an edit control, where the variable of the first TVL column is defined by the first row, each variable is defined in its own row, and the row order in the dialog window corresponds to the column order of the TVL to be created. The name of a variable may consist of up to 12 characters; the first character cannot be a digit. Finally a TVL which has no row as yet is created after pushing the button labeled by Create.

There are two methods to define the rows of the TVL. The first one uses the possibility to append ternary vectors to a selected TVL. Appending all the required vectors to an existing empty TVL leads to the complete TVL. In order to do this, the item Append Ternary Vector(s)... in the menu Objects must be activated, and this opens the dialog window shown in Fig. 1.5. The TVL which is supposed to be extended must be selected in this dialog window, and the respective ternary vectors must be written in an edit control, row by row. It is possible to change these rows. Pushing the OK-button starts the action which appends the defined rows to the selected TVL. This finishes the first method to define a TVL.

The second method to define the rows of a TVL uses the editor which is available in the Single View and in each area of the 4-fold View. If a TVL is shown in such a view, a double-

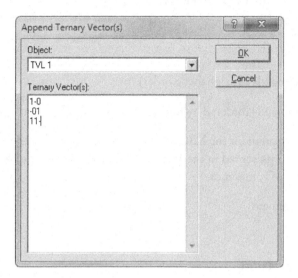

Figure 1.5: Dialog window to append ternary vectors to a selected TVL.

click inside the view switches between the TVL mode that displays the TVL and the edit mode that allows us to change given ternary values and to add ternary vectors to a TVL. The element to be changed can be selected by a mouse click on the position of the element. After the change or an input operation the cursor moves to the next element, first within the row and at the end of one row to the beginning of the next row. This movement of the cursor is controlled by the XBOOLE Monitor. For the input of a TVL only the keys 0, 1, and – are required. A double-click finishes the edit mode. Alternatively, a right click in the view and the appropriate selection in the opened context menu can be used, a very useful approach which makes all the required input operations very feasible.

In order to control the actions of the XBOOLE Monitor, a command language has been defined. All the information about a particular action to be executed will be included into the command; therefore, no additional dialog windows are required. This speeds up the handling of the XBOOLE Monitor in comparison to the toolbars even more. Of course, the application of commands to control the actions of the XBOOLE Monitor requires the knowledge of the syntax of the command language.

In order to study the details of all commands, it is suggested to use the help environment that can be reached by the item Help Topics of the menu labeled '?'. In the contents tab there is an item Command Line that includes the following items:

- List of the Commands

- Topics

- Index of Commands (in alphabetic order)

- Index of Commands (ordered by topics).

A good starting point for the XBOOLE command language is the List of the Commands. Similar information is presented in the Index of Commands (ordered by topics). There are four main categories of commands:

- Object Management

- Derivatives

- Operations for Matrices

- Operations for Sets.

Actions that can be activated by items of the menus File and Objects can be activated alternatively by commands of the category Object Management. There is a complete association of the commands of the category Derivatives with the items of the menu Derivatives. The same statement is true for Matrices and Sets. The menus of the last three categories help the user to learn the XBOOLE command language, because the keyword of the command is written before the associated item text in the menu.

The XBOOLE command language is used in several implementations of XBOOLE Monitors. The XBOOLE Monitor used in addition to this book does not support all commands. These commands which are not supported are listed in a special section of the XBOOLE-Monitor! Index of Commands (ordered by topics). If such a command is used in the command line, it does not cause an error, but it does not activate any action. An example for such a command is help, which can simply be activated using the button F1.

Vice versa, the XBOOLE Monitor used in addition to this book supports several commands which have been defined in addition to the XBOOLE command language. The keywords of these commands begin with an underline character. These additionally supported commands are listed in the section Extended Operations of the index of commands ordered by topics. Most of these commands have a command in the XBOOLE command language which is more or less equivalent. They differ only in the use of VT: the VT is explicitly given in the command, not by an object number.

All XBOOLE commands have the same structure. The keyword is followed by parameters which specify the objects to be manipulated by this command. Based on predefined default values, some of the parameters can be omitted. If the command requires larger additional information, it must be given in consecutive lines and finished by a dot (.) on the same or next line.

One example is the command for the input of a VT that has the following syntax:

```
vtin [ sni [ vtno ] ]
var_list
```

The keyword is vtin. The parameter sni means "space number input" and specifies the Boolean space to which the VT must be assigned while the parameter vtno means "VT number output" and specifies the object number for accessing this VT in the future. On the following lines the ordered list of variables (var_list) must be given. In the lines the variables are separated by space characters. The end of the list is indicated by a dot. The brackets indicate that the parameters are optional. If vtno is not specified the next free object number will be used. The default value for sni is equal to 1. If only one parameter is used, it must be sni. The command for input of the VT $< b \ c >$ in Boolean space 1 as object number 2 is given as follows:

```
vtin 1 2
b c.
```

Practically, you have to type vtin 1 2 and press ENTER. Then the command line will be empty again, and you type b c. and press ENTER again. Then the object will be created and can be seen in one of the views.

An intelligent help system supports the user during the typing of the commands in the command line. Starting with the first letter, all fitting keywords of valid commands are shown in a small help window. After typing the complete keyword, the required parameter structure of the command is shown in this window. If the command needs further information in consecutive lines, the required syntax will be explained in this help window during the input. In this way, it is easy to learn the command language.

1.7.3   XBOOLE PROBLEM PROGRAM

More complex tasks require a sequence of XBOOLE actions for calculating the solution. In order to achieve this, all three possibilities for controlling XBOOLE can be used together. If in such a sequence an error occurs, certain parts of the sequence must be repeated. In such a situation XBOOLE problem programs (PRP) are a valuable support. Instead of activating the calculation step by step, a sequence of commands is written directly into a text file using any standard editor. Such a PRP can be executed by the XBOOLE Monitor step by step or completely without breaks. Such a PRP is also very helpful when the same sequence of XBOOLE actions must be executed several times for data which have completely or partially changed.

The content of a PRP is a sequence of XBOOLE commands. Any file name is allowed for a PRP, and the extension .prp is suggested.

A PRP can be created outside of the XBOOLE Monitor using any text editor or by the XBOOLE Monitor itself. As mentioned above, the XBOOLE commands of all actions executed by the XBOOLE Monitor will be included into a protocol, independent on the way of their specification.

This protocol can be stored as a PRP file using the item `Save protocol as PRP...` in the menu `Extras`. In order to prepare several PRP files in this way, the old protocol can be deleted using the item `Delete Protocol` in the menu `Extras`.

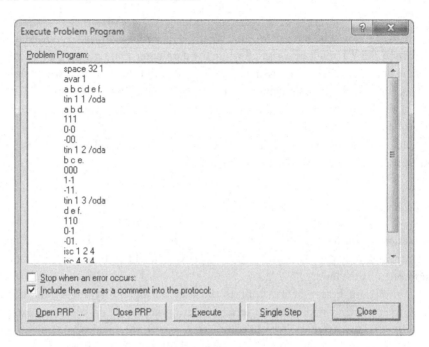

Figure 1.6: Dialog window for the execution of problem programs.

The execution of a PRP can be activated by the item `Execute PRP...` in the menu `Extras` or by the icon `PRP` in the toolbar `Extras`. In both cases a dialog window with the name `Execute Problem Program` appears. A click on the button `Open PRP...` opens a file dialog window where the PRP file can be selected. The content of the PRP file will be shown in the dialog window *Execute Problem Program*. Fig. 1.6 shows this dialog window that includes a loaded PRP. Using the scroll bar the other commands of the PRP can be visualized. A click on the button `Single Step` executes a single command beginning with the first command and then according to the sequence of the PRP. A click on the button `Execute` executes the sequence of commands completely.

As an example, we calculate the behavior of a logic circuit. Fig. 1.7 depicts the structure of the circuit (a) and lists the PRP that calculates the complete behavior in a minimized form (b). The first command in the PRP defines a Boolean space 1 of 32 variables. The second command `avar` associates the variables a, b, c, d, e, and f in this order to the Boolean space 1. In this way a well ordered output is organized. Thereafter the input of the phase lists of the three gates is described by three commands `tin`. Each TVL is associated with the Boolean space 1. The object numbers of these TVLs correspond to the labels of the gates in Fig. 1.7 (a). The behavior is calculated by the

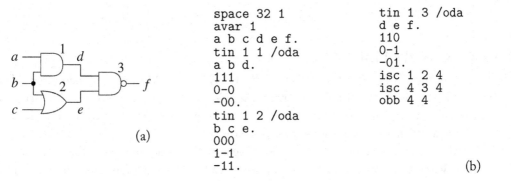

```
space 32 1          tin 1 3 /oda
avar 1              d e f.
a b c d e f.        110
tin 1 1 /oda        0-1
a b d.              -01.
111                 isc 1 2 4
0-0                 isc 4 3 4
-00.                obb 4 4
tin 1 2 /oda
b c e.
000
1-1
-11.
```

(a)

(b)

Figure 1.7: Structure of a simple circuit (a) and the associated problem program to calculate the complete behavior (b).

intersection of all three phase lists and stored as object number 4. Finally, by means of an orthogonal block building using the obb command, the resulting TVL is minimized with respect to the number of rows without change of the object number. The loaded PRP is partially shown in Fig. 1.6.

Figure 1.8: The complete behavior calculated by means of a PRP listed in Fig. 1.7.

Fig. 1.8 shows in the 4-fold View all basic TVLs and the TVL of the desired complete behavior. A comparison with the PRP of Fig. 1.7 (b) reveals that the input TVLs are orthogonalized in the input action because the form oda was specified. An orthogonal form is a precondition for the intersection operation of XBOOLE. The result in the right bottom part of the 4-fold View shows all behavior details of the circuit of Fig. 1.7 (a). The output f is equal to 0 if the inputs a, b, and d are equal to 1 for each value of the input c. In all other cases the output f is equal to 1.

We use PRPs in Sec. 3 for several approaches to solve Boolean differential equations.

1.7.4   XBOOLE LIBRARY

The basis of the XBOOLE Monitor is the XBOOLE Library. Most of the functions of this library are wrapped by the XBOOLE Monitor for a simple handling. The XBOOLE Library is written in the programming language C and can be used in programs written in programming languages C, C++, Java, and other languages on several platforms.

It is necessary to know many programming details in order to use the XBOOLE Library. In this book we focus on approaches to solve Boolean differential equations and not on programming aspects. Therefore the XBOOLE Library will not be considered in the remaining chapters although their functions can be used in a program that solves Boolean differential equations. The XBOOLE Library can be ordered using the following address:

Steinbeis-TZ Logische Systeme
Nelkentor 7
D-09126 Chemnitz
Germany
FAX: +49 371 5381 929
Email: SU0158@stw.de.

By using a simple example, the application of the XBOOLE Library will be indicated here. Assume there is a TVL in which k variables occur. It may be so that the associated Boolean function actually depends only on l variables, where $l < k$. It is the task of the function simplify, to create the TVL of a new simplified function which only includes such variables on which the function really depends.

```
#include "xb_port.h"

void simplify(uns **ti, uns **res)
/* *ti pointer to the logic function to be simplified */
/* *res pointer to the simplified function */
{
  uns *xi;                        /* selected variable */
  xi = NULL;                      /* initialize xi */
  COPYOBJ(ti, res);               /* copy ti to res */
  while(SV_NEXT(ti, &xi, &xi))    /* select variable */
    if(TE_DERK(res, &xi))         /* if res does not depend on xi */
      MAXK(res, &xi, res);        /* remove xi from res */
  OBB(res, res);                  /* minimize rows */
}
```

Figure 1.9: C-program to simplify a logic function using the XBOOLE Library.

Fig. 1.9 shows the source code of a C-function that removes all dependent variables from a TVL and shortens the number of rows in the result. The prototypes of all functions of the XBOOLE Library are defined in the header file "xb_port.h". In order to use the XBOOLE Library, this header file must be included into the source file. All XBOOLE objects are managed by XBOOLE, and the access is possible by pointers to the type uns that is defined in the XBOOLE header too.

The function simplify gets access to the given TVL by the parameter ti and returns the simplified TVL by the parameter res. This function requires that an orthogonal TVL is given and ensures that the form of the TVL does not change. All XBOOLE functions are indicated by capital letters. The function COPYOBJ copies the TVL ti to a new TVL res such that the given TVL will not change. In order to facilitate the access to each variable in ti separately, a TVL xi is used. After its definition and initialization the XBOOLE function SV_NEXT assigns in the first sweep of the loop the first variable of the TVL ti to the TVL xi, and in the following sweeps the next variables, one after the other, respectively. The XBOOLE function SV_NEXT returns the value false if no further variable exists in the TVL ti.

In the loop the XBOOLE function TE_DERK checks whether the derivative of ti with regard to xi is equal to 0. If that is true, the function ti does not depend on xi, and xi will be removed from the TVL res by means of a k-fold maximum operation realized by the XBOOLE function MAXK. The XBOOLE function OBB finally reduces the number of rows in the simplified orthogonal TVL res.

SUMMARY

In this chapter we have discussed very briefly some important terms of the Boolean Algebra. The Boolean spaces B^n are built from the elements 0 and 1 or vectors of these elements. Boolean functions are unique mappings from B^n into the simplest Boolean space B. Boolean expressions of variables and operations describe in a compact manner Boolean functions. A Boolean equation is built from two Boolean functions and has a set of Boolean vectors as result. The Boolean Algebra is widely used to specify, design, analyze, and test digital circuits and systems.

EXERCISES

1.1 De Morgan's Law is valid for the operations $\wedge$ and $\vee$. Holds a similar law also for the operations $\odot$ and $\oplus$ in: $\overline{x_1 \odot x_2} = \overline{x}_1 \oplus \overline{x}_2$?

1.2 The formula $F_1 = \overline{a}\,\overline{b} \vee (b \oplus a\,b\,\overline{c}) \vee \overline{b}\,c$ contains the Boolean variables a, b, c; the formula $F_2 = (a\,\overline{c}\,\overline{d} \oplus \overline{d}) \vee \overline{a}\,d \vee c\,d$ contains the Boolean variables a, c, and d. Do these formulas describe the same function: $f(F_1) = f(F_2)$?

1.3 Which binary vectors belong to the solution of the Boolean equation:

$$a\,b \oplus \overline{c} = b \oplus \overline{a}\,c \oplus \overline{a}\,\overline{b}\,\overline{c}\,. \tag{1.28}$$

Is the equation (1.28) solvable with regard to each of the three variables? If YES, all functions $a = f_1(b, c), b = f_2(a, c)$, and $c = f_3(a, b)$ are searched.

1.4 Follow the hints for the download of Subsec. 1.7.2 to prepare a usable XBOOLE Monitor on your computer. Study the XBOOLE help information using the menu item '?/Help Topics'. Hint: in Windows 7 the file WinHlp32.exe, which is available on the web, must be installed in order to read the file xbm32.hlp. Define a Boolean space and the TVL:

$$XBO[1] = \begin{array}{cccc} a & b & c & d \\ \hline 0 & - & 1 & 1 \\ 1 & 1 & 1 & 0 \\ - & 1 & 0 & - \\ \hline \end{array} \qquad (1.29)$$

as XBOOLE object 1. Use the alternative XBOOLE commands cco and _cco in order to create the XBOOLE object 2 in which the content of the first two columns is exchanged with the content of the last two columns ($a \leftrightarrow c, b \leftrightarrow d$) of $XBO[1]$ (1.29).

1.5 Prepare a PRP file as shown in Fig. 1.7 (b). Execute this PRP in the XBOOLE Monitor. Note: this PRP can be executed in a new XBOOLE Monitor window, without a precondition, because the space definition is included in the PRP. Show all the created TVLs in the *4-fold View* and verify the result using Fig. 1.8.

CHAPTER 2

Summary of the Boolean Differential Calculus

2.1 INTRODUCTION

The development of the Boolean Differential Calculus was initialized by the necessity to detect errors in digital circuits in the 1950s. It was developed since the 1970s based on the original ideas of Reed [1954], Huffman [1958] and Akers [1959]. In these early papers the term *Boolean difference* was used for the EXOR of two subfunctions of a logic function. Generally, the finite set B^n with $\oplus$ as *addition* and *difference* and $\wedge$ as multiplication satisfies the axioms of a *Boolean ring*.

One comprehensive presentation were the books of Thayse and his colleagues Davio, Deschamps and Thayse [1978], Thayse [1981], among others. A larger research center dealing with these theories and applications was the Chemnitz University of Technology where the authors are also coming from. A monograph Bochmann and Posthoff [1981] of this calculus was published in German language. A comprehensive presentation was written as part of Posthoff and Steinbach [2004].

It is also very important to mention, particularly in the sense of applications that in parallel the comprehensive software package XBOOLE Steinbach [1992] with several versions has been developed (mainly by B. Steinbach and his research group) where the operations and theorems of the Boolean Differential Calculus are implemented in such a way that they can be used in different areas. This is very important because it restricts the efforts of researchers and applicants to the modeling of the problem, the software *is doing the rest of the work* (as usual). Complementary to the textbook Posthoff and Steinbach [2004] now the book Steinbach and Posthoff [2009] is available that contains many examples and exercises together with their solutions. The Boolean Differential Calculus together with XBOOLE on a standard computer allow us to solve the given real-life problems on a high level of abstraction in a very efficient way.

The term *Boolean Differential Calculus* may generate the demand for a *Boolean Integral Calculus*. In such an inverse calculus, the original logic functions are wanted when the results of any differential operations are known. This question was shortly discussed in the monograph Bochmann and Posthoff [1981]. A comprehensive presentation of the Boolean Integral Calculus is given in the Ph.D. thesis of Steinbach [1981] and allows to express and to handle sets of logic functions. The main ideas of this Boolean Integral Calculus which show how a Boolean differential equation can be solved were summarized in one part of Posthoff and Steinbach [2004].

There are also very interesting extensions of the Boolean Differential Calculus to multi-valued logics Yanushkevich [1998]. Two chapters of the monograph (Bochmann and Posthoff [1981]) explore the differential calculus for combined Boolean and multi-valued logics. Some remarks about this approach are repeated in Bochmann [2008]. Here we restrict ourselves to the Differential Calculus in the Boolean domain.

2.2   SIMPLE DERIVATIVE OPERATIONS

When the value of a binary variable x is considered, then it can only change from 0 to 1 or from 1 to 0, and these two possibilities can be combined by the change from x to $\overline{x}$. Now we want to see the influence of this change on the behavior of a function $f(\mathbf{x})$. In order to do this we use the *antivalence*.

Definition 2.1   Let $f(\mathbf{x}) = f(x_1, \ldots, x_i, \ldots, x_n)$ be a Boolean function of n variables, then

$$\frac{\partial f(\mathbf{x})}{\partial x_i} = f(x_1, \ldots, x_i, \ldots, x_n) \oplus f(x_1, \ldots, \overline{x}_i, \ldots, x_n) \qquad (2.1)$$

is the (*simple*) *derivative* of the Boolean function $f(\mathbf{x})$ with regard to the variable x_i.

It follows from (2.1) that the derivative $\frac{\partial f(\mathbf{x})}{\partial x_i}$ is a Boolean function that is equal to 1 if the function value of $f(x_1, \ldots, x_i, \ldots, x_n)$ differs from the function value after the change of x_i, expressed by $f(x_1, \ldots, \overline{x}_i, \ldots, x_n)$. In other words, the derivative of $f(\mathbf{x})$ with regard to x_i describes whether or not the change of the variable x_i causes a change of the function value.

Example 2.2   Let $f(a, b, c) = \overline{a}\,\overline{b}\,\overline{c} \oplus a\,b$ be a given Boolean function. Based on definition (2.1), the simple derivative $\frac{\partial f(a,b,c)}{\partial c}$ can be calculated as follows:

$$\begin{aligned}
\frac{\partial f(a, b, c)}{\partial c} &= f(a, b, c) \oplus f(a, b, \overline{c}) \\
&= (\overline{a}\,\overline{b}\,\overline{c} \oplus a\,b) \oplus (\overline{a}\,\overline{b}\,c \oplus a\,b) \\
&= \overline{a}\,\overline{b}\,(\overline{c} \oplus c) \\
&= \overline{a}\,\overline{b}\,.
\end{aligned}$$

The function $f(a, b, c)$ changes its value for $a = 0$ and $b = 0$: $f(0, 0, 0) \neq f(0, 0, 1)$. This can be checked easily because $f(0, 0, 0) = 1$ and $f(0, 0, 1) = 0$.

The derivative of $f(a, b, c)$ with regard to c does not depend on c anymore as can be seen from the following theorem.

Theorem 2.3   *It holds for* $f(\mathbf{x}) = f(x_1, \ldots, x_i, \ldots, x_n)$ *that*

$$\frac{\partial f(\mathbf{x})}{\partial x_i} = f(x_1, \ldots, x_i = 0, \ldots, x_n) \oplus f(x_1, \ldots, x_i = 1, \ldots, x_n). \qquad (2.2)$$

Proof. We use the understanding that the transition $x \rightarrow \overline{x}$ is either a transition $0 \rightarrow 1$ or a transition $1 \rightarrow 0$. Because of $1 \oplus 0 = 0 \oplus 1$ we can always use the given theorem. The direct insertion of 0 and 1 very often simplifies the calculations. □

The derivative can be used to determine whether a function $f(\mathbf{x})$ does not depend on a variable x_i. Recall, the function f is independent on the variable x_i if $f(x_1, \ldots, x_i = 0, \ldots, x_n) = f(x_1, \ldots, x_i = 1, \ldots, x_n)$. If this is the case then we have immediately $f(x_1, \ldots, x_i = 0, \ldots, x_n) \oplus f(x_1, \ldots, x_i = 1, \ldots, x_n) = 0$, i.e., $\frac{\partial f(\mathbf{x})}{\partial x_i} = 0$. If, vice versa, $\frac{\partial f(\mathbf{x})}{\partial x_i} = 0$ holds, we get $f(x_i = 0) \oplus f(x_i = 1) = 0$, i.e., $f(x_i = 0) = f(x_i = 1)$.

The replacement of $\oplus$ by $\wedge$ leads to the definition of the *simple minimum*.

Definition 2.4   Let $f(\mathbf{x}) = f(x_1, \ldots, x_i, \ldots, x_n)$ be a Boolean function of n variables, then

$$\min_{x_i} f(\mathbf{x}) = f(x_1, \ldots, x_i, \ldots, x_n) \wedge f(x_1, \ldots, \overline{x}_i, \ldots, x_n) \tag{2.3}$$

is the *(simple) minimum* of the Boolean function $f(\mathbf{x})$ with regard to the variable x_i.

The minimum of $f(\mathbf{x})$ is less than or equal to the Boolean function $f(\mathbf{x})$ because $f(\mathbf{x})$ occurs as part of an $\wedge$ - operation in Def. (2.3). Therefore this derivative operation is called *minimum*.

Example 2.5   We take the function $f(a, b, c) = \overline{a}\,\overline{b}\,\overline{c} \oplus a\,b$ from the previous example and calculate the simple minimum:

$$\min_{c} f(a, b, c) = f(a, b, c) \wedge f(a, b, \overline{c})$$
$$= (\overline{a}\,\overline{b}\,\overline{c} \oplus a\,b) \wedge (\overline{a}\,\overline{b}\,c \oplus a\,b)$$
$$= a\,b \ .$$

Again the result does not depend on the variable c.

Theorem 2.6   *It holds for* $f(\mathbf{x}) = f(x_1, \ldots, x_i, \ldots, x_n)$ *that*

$$\min_{x_i} f(\mathbf{x}) = f(x_1, \ldots, x_i = 0, \ldots, x_n) \wedge f(x_1, \ldots, x_i = 1, \ldots, x_n). \tag{2.4}$$

The proof follows the same idea that has been used for the derivative. Finally, we define the *simple maximum*.

Definition 2.7   Let $f(\mathbf{x}) = f(x_1, \ldots, x_i, \ldots, x_n)$ be a Boolean function of n variables, then

$$\max_{x_i} f(\mathbf{x}) = f(x_1, \ldots, x_i, \ldots, x_n) \vee f(x_1, \ldots, \overline{x}_i, \ldots, x_n) \tag{2.5}$$

is the *(simple) maximum* of the Boolean function $f(\mathbf{x})$ with regard to the variable x_i.

The simple maximum $\max_{x_i} f(\mathbf{x})$ is a Boolean function that is equal to 1 if the function value 1 of $f(\mathbf{x})$ appears at least once for the two values of x_i. The name *maximum* emphasizes that the result of this derivative operation is larger than or equal to the basic function $f(\mathbf{x})$ because $f(\mathbf{x})$ occurs as part of an $\vee$ - operation in Def. (2.5).

Example 2.8   We take again the function $f(a, b, c) = \overline{a}\,\overline{b}\,\overline{c} \oplus a\,b$ from the previous examples, transform it into $f(a, b, c) = \overline{a}\,\overline{b}\,\overline{c} \vee a\,b$, and calculate the simple maximum $\max_c f(a, b, c)$:

$$
\begin{aligned}
\max_c f(a, b, c) &= f(a, b, c) \vee f(a, b, \overline{c}) \\
&= (\overline{a}\,\overline{b}\,\overline{c} \vee a\,b) \vee (\overline{a}\,\overline{b}\,c \vee a\,b) \\
&= \overline{a}\,\overline{b}\,(\overline{c} \vee c) \vee a\,b \\
&= \overline{a}\,\overline{b} \vee a\,b.
\end{aligned}
$$

The next theorem holds in the same way as before.

Theorem 2.9   *It holds for $f(\mathbf{x}) = f(x_1, \ldots, x_i, \ldots, x_n)$ of n variables that*

$$
\max_{x_i} f(\mathbf{x}) = f(x_1, \ldots, x_i = 0, \ldots, x_n) \vee f(x_1, \ldots, x_i = 1, \ldots, x_n). \tag{2.6}
$$

There are several relations between the simple derivatives. Figure 2.1 summarizes the results of the previous three examples. The small arrows within the Karnaugh-map of $f(a, b, c)$ indicate the pairs of function values that cause the calculated results of the derivative operations with regard to c.

The Karnaugh-maps in the middle row of Figure 2.1 show the results of the derivative operations calculated by Def. (2.3), (2.1), and (2.5), respectively. These Karnaugh-maps confirm that simple derivative operations do not depend on the variable that has been selected for the calculation.

The Karnaugh-maps in the lowest row of Figure 2.1 show the results of the derivative operations calculated by Th. (2.4), (2.2), and (2.6), respectively. They originate also from the Karnaugh-maps in the middle row of Figure 2.1 by reduction with regard to c. The comparison of these Karnaugh-maps confirms that the antivalence of all three simple derivative operations is equal to 0 (2.7). This rule can be verified by substitution of definition (2.1), (2.3), and (2.5) into (2.7) and the simplification of the resulting expression.

$$
\min_{x_i} f(\mathbf{x}) \oplus \frac{\partial f(\mathbf{x})}{\partial x_i} \oplus \max_{x_i} f(\mathbf{x}) = 0. \tag{2.7}
$$

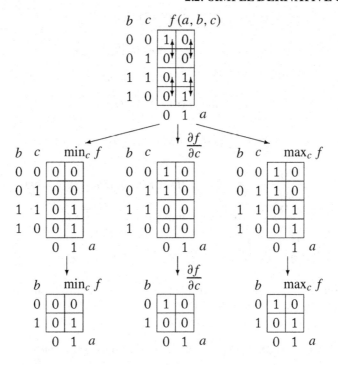

Figure 2.1: The Karnaugh-maps of $f(a, b, c) = \bar{a}\,\bar{b}\,\bar{c} \oplus a\,b$ and all simple derivatives with regard to c.

From (2.7) follow three rules to calculate a simple derivative operation using the other two simple derivative operations:

$$\frac{\partial f(\mathbf{x})}{\partial x_i} = \min_{x_i} f(\mathbf{x}) \oplus \max_{x_i} f(\mathbf{x}) , \tag{2.8}$$

$$\min_{x_i} f(\mathbf{x}) = \frac{\partial f(\mathbf{x})}{\partial x_i} \oplus \max_{x_i} f(\mathbf{x}) , \tag{2.9}$$

$$\max_{x_i} f(\mathbf{x}) = \min_{x_i} f(\mathbf{x}) \oplus \frac{\partial f(\mathbf{x})}{\partial x_i} . \tag{2.10}$$

The order relation (2.11) between the minimum $\min_{x_i} f(\mathbf{x})$, the Boolean function $f(\mathbf{x})$ and the maximum $\max_{x_i} f(\mathbf{x})$ can be confirmed by comparing the outer Karnaugh-maps in the middle row and the Karnaugh-map on top of Figure 2.1. It is easy to verify that the equality in (2.11) holds if $f(\mathbf{x})$ is independent on x_i:

$$\min_{x_i} f(\mathbf{x}) \leq f(\mathbf{x}) \leq \max_{x_i} f(\mathbf{x}) . \tag{2.11}$$

From (2.11) follow three Eqs. (2.12), (2.13), and (2.14):

$$\min_{x_i} f(\mathbf{x}) \wedge \overline{f(\mathbf{x})} = 0 , \tag{2.12}$$

$$f(\mathbf{x}) \wedge \overline{\max_{x_i} f(\mathbf{x})} = 0 , \tag{2.13}$$

$$\min_{x_i} f(\mathbf{x}) \wedge \overline{\max_{x_i} f(\mathbf{x})} = 0 . \tag{2.14}$$

Using (2.8) and (2.14) we get the constraint (2.15):

$$\frac{\partial f(\mathbf{x})}{\partial x_i} \wedge \overline{\max_{x_i} f(\mathbf{x})} = (\min_{x_i} f(\mathbf{x}) \oplus \max_{x_i} f(\mathbf{x})) \wedge \overline{\max_{x_i} f(\mathbf{x})} ,$$

$$\frac{\partial f(\mathbf{x})}{\partial x_i} \wedge \overline{\max_{x_i} f(\mathbf{x})} = \min_{x_i} f(\mathbf{x}) \wedge \overline{\max_{x_i} f(\mathbf{x})} ,$$

$$\frac{\partial f(\mathbf{x})}{\partial x_i} \wedge \overline{\max_{x_i} f(\mathbf{x})} = 0 . \tag{2.15}$$

Based on (2.8) and (2.14) it can be shown that the simple minimum is orthogonal to the simple derivative (2.16):

$$\min_{x_i} f(\mathbf{x}) \wedge \frac{\partial f(\mathbf{x})}{\partial x_i} = \min_{x_i} f(\mathbf{x}) \wedge (\min_{x_i} f(\mathbf{x}) \oplus \max_{x_i} f(\mathbf{x})) ,$$

$$\min_{x_i} f(\mathbf{x}) \wedge \frac{\partial f(\mathbf{x})}{\partial x_i} = \min_{x_i} f(\mathbf{x}) \oplus (\min_{x_i} f(\mathbf{x}) \wedge \max_{x_i} f(\mathbf{x})) ,$$

$$\min_{x_i} f(\mathbf{x}) \wedge \frac{\partial f(\mathbf{x})}{\partial x_i} = \min_{x_i} f(\mathbf{x}) \wedge (1 \oplus \max_{x_i} f(\mathbf{x})) ,$$

$$\min_{x_i} f(\mathbf{x}) \wedge \frac{\partial f(\mathbf{x})}{\partial x_i} = \min_{x_i} f(\mathbf{x}) \wedge \overline{\max_{x_i} f(\mathbf{x})} ,$$

$$\min_{x_i} f(\mathbf{x}) \wedge \frac{\partial f(\mathbf{x})}{\partial x_i} = 0 . \tag{2.16}$$

The formulas (2.8), (2.9), and (2.10) can be transformed into alternative rules to calculate a simple derivative operation using the other two simple derivative operations. Formula (2.17) can be based on (2.8) and takes advantage from (2.14). The simple minimum (2.18) follows from (2.9) and (2.15). Finally, the simple maximum (2.19) is created from (2.10) and uses the orthogonality (2.16) between simple minimum and the simple derivative:

$$\frac{\partial f(\mathbf{x})}{\partial x_i} = \max_{x_i} f(\mathbf{x}) \wedge \overline{\min_{x_i} f(\mathbf{x})} , \tag{2.17}$$

$$\min_{x_i} f(\mathbf{x}) = \max_{x_i} f(\mathbf{x}) \wedge \overline{\frac{\partial f(\mathbf{x})}{\partial x_i}} , \tag{2.18}$$

$$\max_{x_i} f(\mathbf{x}) = \min_{x_i} f(\mathbf{x}) \vee \frac{\partial f(\mathbf{x})}{\partial x_i} . \tag{2.19}$$

2.3   VECTORIAL DERIVATIVES

The change of values is not restricted to a single variable. *Vectorial derivatives* describe the properties of the Boolean function $f(\mathbf{x})$ if several of its variables change the values simultaneously.

Definition 2.10   Let $\mathbf{x}_0 = (x_1, x_2, ..., x_k), \mathbf{x}_1 = (x_{k+1}, x_{k+2}, ..., x_n)$ be two disjoint sets of Boolean variables, and $f(\mathbf{x}_0, \mathbf{x}_1) = f(x_1, x_2, ..., x_n) = f(\mathbf{x})$ a Boolean function of n variables, then

$$\frac{\partial f(\mathbf{x}_0, \mathbf{x}_1)}{\partial \mathbf{x}_0} = f(\mathbf{x}_0, \mathbf{x}_1) \oplus f(\overline{\mathbf{x}}_0, \mathbf{x}_1) \tag{2.20}$$

is the *vectorial derivative* of the Boolean function $f(\mathbf{x}_0, \mathbf{x}_1)$ with regard to the variables of $\mathbf{x}_0$.

The vectorial derivative $\frac{\partial f(\mathbf{x}_0, \mathbf{x}_1)}{\partial \mathbf{x}_0}$ is a logic function that depends generally on all variables $\mathbf{x} = (\mathbf{x}_0, \mathbf{x}_1)$. A function value 1 of the vectorial derivative $\frac{\partial f(\mathbf{x}_0, \mathbf{x}_1)}{\partial \mathbf{x}_0}$ indicates that the corresponding value of $f(\mathbf{x}_0, \mathbf{x}_1)$ will change if all variables of $\mathbf{x}_0$ change their value simultaneously.

Example 2.11   Let $f(a, b, c) = \overline{a} b \overline{c} \oplus a c$ be a given Boolean function. Based on Def. (2.20) the vectorial derivative $\frac{\partial f(a,b,c)}{\partial (a,c)}$ can be calculated as follows:

$$\begin{aligned}
\frac{\partial f(a, b, c)}{\partial (a, c)} &= f(a, b, c) \oplus f(\overline{a}, b, \overline{c}) \\
&= (\overline{a} b \overline{c} \oplus a c) \oplus (a b c \oplus \overline{a}\,\overline{c}) \\
&= \overline{a}\,\overline{c}\,(b \oplus 1) \oplus a c\,(b \oplus 1) \\
&= \overline{b}\,(\overline{a}\,\overline{c} \oplus a c) \\
&= \overline{b}\,(\overline{a} \oplus c)\,.
\end{aligned}$$

The example shows that the vectorial derivative depends on all the variables of the given function.

Again, the $\oplus$ - operation of the vectorial derivative can be replaced by one of the non-linear operations.

Definition 2.12   Let $\mathbf{x}_0 = (x_1, x_2, ..., x_k), \mathbf{x}_1 = (x_{k+1}, x_{k+2}, ..., x_n)$ be two disjoint sets of Boolean variables, and $f(\mathbf{x}_0, \mathbf{x}_1) = f(x_1, x_2, ..., x_n) = f(\mathbf{x})$ a Boolean function of n variables, then

$$\min_{\mathbf{x}_0} f(\mathbf{x}_0, \mathbf{x}_1) = f(\mathbf{x}_0, \mathbf{x}_1) \wedge f(\overline{\mathbf{x}}_0, \mathbf{x}_1) \tag{2.21}$$

is the *vectorial minimum*, and

$$\max_{\mathbf{x}_0} f(\mathbf{x}_0, \mathbf{x}_1) = f(\mathbf{x}_0, \mathbf{x}_1) \vee f(\overline{\mathbf{x}}_0, \mathbf{x}_1) \tag{2.22}$$

is the *vectorial maximum* of the Boolean function $f(\mathbf{x}_0, \mathbf{x}_1)$ with regard to the variables of $\mathbf{x}_0$.

Generally, the vectorial minimum (2.21) and the vectorial maximum (2.22) depend on all variables.

A function value 1 occurs in the vectorial minimum $\min_{\mathbf{x}_0} f(\mathbf{x}_0, \mathbf{x}_1)$ if the corresponding value of $f(\mathbf{x}_0, \mathbf{x}_1)$ remains equal to 1 when all the variables of $\mathbf{x}_0$ change their value simultaneously. Because of this restriction the vectorial minimum is less than or equal to the function $f(\mathbf{x}_0, \mathbf{x}_1)$. Therefore the name *vectorial minimum* was chosen.

The comparison of (2.21) and (2.22) shows that the same pair of function values $f(\mathbf{x}_0, \mathbf{x}_1)$ determines the function value of the vectorial maximum. It is sufficient that at least one of them is equal to 1 in order get the function value 1 of $\max_{\mathbf{x}_0} f(\mathbf{x}_0, \mathbf{x}_1)$. Since $f(\mathbf{x}_0, \mathbf{x}_1)$ occurs as part of an $\vee$ - operation in definition (2.22) the vectorial maximum is larger than or equal to the corresponding function $f(\mathbf{x}_0, \mathbf{x}_1)$. This explains the name *vectorial maximum* for this derivative operation.

Example 2.13   For an easy comparison with the vectorial derivative, we take the same function $f(a, b, c) = \overline{a}\, b\, \overline{c} \oplus a\, c$, and calculate first the vectorial minimum $\min_{(a,c)} f(a, b, c)$ based on Def. (2.21) and thereafter the vectorial maximum $\max_{(a,c)} f(a, b, c)$ based on Def. (2.22):

$$\min_{(a,c)} f(a, b, c) = f(a, b, c) \wedge f(\overline{a}, b, \overline{c})$$
$$= (\overline{a}\, b\, \overline{c} \oplus a\, c) \wedge (a\, b\, c \oplus \overline{a}\, \overline{c})$$
$$= (\overline{a}\, b\, \overline{c} \oplus a\, b\, c)$$
$$= b\,(\overline{a}\, \overline{c} \oplus a\, c)$$
$$= b\,(\overline{a} \oplus c)\,,$$

$$\max_{(a,c)} f(a, b, c) = f(a, b, c) \vee f(\overline{a}, b, \overline{c})$$
$$= (\overline{a}\, b\, \overline{c} \oplus a\, c) \vee (a\, b\, c \oplus \overline{a}\, \overline{c})$$
$$= (\overline{a}\, b\, \overline{c} \vee a\, c) \vee (a\, b\, c \vee \overline{a}\, \overline{c})$$
$$= a\, c \vee \overline{a}\, \overline{c}$$
$$= \overline{a} \oplus c\,.$$

Generally the result of a vectorial operation (derivative, minimum, maximum) depends on all the variables $x_1, \ldots, x_n$. For special functions, however, exceptions are possible.

Relations between the vectorial derivatives can easily be seen in Fig. 2.2 which summarizes the examples of the vectorial derivative operations. The small arrows in the Karnaugh-map of $f(a, b, c)$ indicate the pairs of function values which cause the calculated results of the derivative operations with regard to (a, c).

The Karnaugh-maps in the second row of Fig. 2.2 show the results of the derivative operations calculated by Def. (2.21), (2.20), and (2.22), respectively. The left two of these Karnaugh-maps show

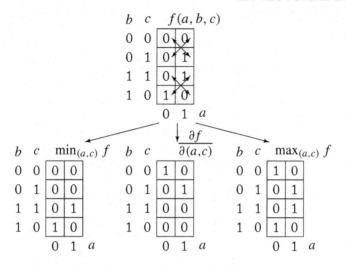

Figure 2.2: The Karnaugh-maps of $f(a, b, c) = \bar{a}\,b\,\bar{c} \oplus a\,c$ and all vectorial derivatives with regard to the variables (a, c).

that the vectorial derivative operations generally depend on all variables. The vectorial maximum in the rightmost Karnaugh-map does not depend on b. A simple example where the vectorial maximum also depends on all variables is $\max_{(a,c)} f(a, b, c)$ of the function $f(a, b, c) = b\,(a \oplus c)$.

The comparison of the outer Karnaugh-maps of the lower row and the Karnaugh-map on top of Fig. 2.2 reveals that the order relation (2.23) holds between the vectorial minimum $\min_{\mathbf{x}_0} f(\mathbf{x}_0, \mathbf{x}_1)$, the Boolean function $f(\mathbf{x}_0, \mathbf{x}_1)$, and the vectorial maximum $\max_{\mathbf{x}_0} f(\mathbf{x}_0, \mathbf{x}_1)$:

$$\min_{\mathbf{x}_0} f(\mathbf{x}_0, \mathbf{x}_1) \leq f(\mathbf{x}_0, \mathbf{x}_1) \leq \max_{\mathbf{x}_0} f(\mathbf{x}_0, \mathbf{x}_1) \,. \tag{2.23}$$

Ineq. (2.23) can be split into the equations:

$$\min_{\mathbf{x}_0} f(\mathbf{x}_0, \mathbf{x}_1) \wedge \overline{f(\mathbf{x}_0, \mathbf{x}_1)} = 0 \,, \tag{2.24}$$

$$f(\mathbf{x}_0, \mathbf{x}_1) \wedge \overline{\max_{\mathbf{x}_0} f(\mathbf{x}_0, \mathbf{x}_1)} = 0 \,, \tag{2.25}$$

$$\min_{\mathbf{x}_0} f(\mathbf{x}_0, \mathbf{x}_1) \wedge \overline{\max_{\mathbf{x}_0} f(\mathbf{x}_0, \mathbf{x}_1)} = 0 \,. \tag{2.26}$$

The Karnaugh-maps in the lower row of Fig. 2.2 show furthermore that the antivalence of all three vectorial derivatives is equal to 0 (2.27), and the vectorial minimum is orthogonal to the vectorial derivative (2.28):

$$\min_{\mathbf{x}_0} f(\mathbf{x}_0, \mathbf{x}_1) \oplus \frac{\partial f(\mathbf{x}_0, \mathbf{x}_1)}{\partial \mathbf{x}_0} \oplus \max_{\mathbf{x}_0} f(\mathbf{x}_0, \mathbf{x}_1) = 0 \,, \tag{2.27}$$

$$\min_{\mathbf{x}_0} f(\mathbf{x}_0, \mathbf{x}_1) \wedge \frac{\partial f(\mathbf{x}_0, \mathbf{x}_1)}{\partial \mathbf{x}_0} = 0 \,. \tag{2.28}$$

Based on (2.27), each vectorial derivative can be calculated using the two other vectorial derivative operations:

$$\min_{\mathbf{x}_0} f(\mathbf{x}_0, \mathbf{x}_1) = \frac{\partial f(\mathbf{x}_0, \mathbf{x}_1)}{\partial \mathbf{x}_0} \oplus \max_{\mathbf{x}_0} f(\mathbf{x}_0, \mathbf{x}_1) , \tag{2.29}$$

$$\frac{\partial f(\mathbf{x}_0, \mathbf{x}_1)}{\partial \mathbf{x}_0} = \min_{\mathbf{x}_0} f(\mathbf{x}_0, \mathbf{x}_1) \oplus \max_{\mathbf{x}_0} f(\mathbf{x}_0, \mathbf{x}_1) , \tag{2.30}$$

$$\max_{\mathbf{x}_0} f(\mathbf{x}_0, \mathbf{x}_1) = \min_{\mathbf{x}_0} f(\mathbf{x}_0, \mathbf{x}_1) \oplus \frac{\partial f(\mathbf{x}_0, \mathbf{x}_1)}{\partial \mathbf{x}_0} . \tag{2.31}$$

Alternative rules follow from (2.23), (2.27), and (2.28):

$$\min_{\mathbf{x}_0} f(\mathbf{x}_0, \mathbf{x}_1) = \max_{\mathbf{x}_0} f(\mathbf{x}_0, \mathbf{x}_1) \wedge \overline{\frac{\partial f(\mathbf{x}_0, \mathbf{x}_1)}{\partial \mathbf{x}_0}} , \tag{2.32}$$

$$\frac{\partial f(\mathbf{x}_0, \mathbf{x}_1)}{\partial \mathbf{x}_0} = \max_{\mathbf{x}_0} f(\mathbf{x}_0, \mathbf{x}_1) \wedge \overline{\min_{\mathbf{x}_0} f(\mathbf{x}_0, \mathbf{x}_1)} , \tag{2.33}$$

$$\max_{\mathbf{x}_0} f(\mathbf{x}_0, \mathbf{x}_1) = \min_{\mathbf{x}_0} f(\mathbf{x}_0, \mathbf{x}_1) \vee \frac{\partial f(\mathbf{x}_0, \mathbf{x}_1)}{\partial \mathbf{x}_0} . \tag{2.34}$$

The simple derivative operations may be considered as a special case of the vectorial derivative operations where the vector $\mathbf{x}_0$ includes only the single variable x_i. With the exception of Th. (2.2), (2.4), and (2.6), comparable properties exist for simple derivative operations and vectorial derivative operations, respectively. Next we list further relations for vectorial derivative operations:

$$\frac{\partial f(\mathbf{x}_0, \mathbf{x}_1)}{\partial \mathbf{x}_0} = \frac{\partial \overline{f}(\mathbf{x}_0, \mathbf{x}_1)}{\partial \mathbf{x}_0} , \tag{2.35}$$

$$\min_{\mathbf{x}_0} f(\mathbf{x}_0, \mathbf{x}_1) = \overline{\max_{\mathbf{x}_0} \overline{f}(\mathbf{x}_0, \mathbf{x}_1)} , \tag{2.36}$$

$$\max_{\mathbf{x}_0} f(\mathbf{x}_0, \mathbf{x}_1) = \overline{\min_{\mathbf{x}_0} \overline{f}(\mathbf{x}_0, \mathbf{x}_1)} , \tag{2.37}$$

$$\frac{\partial (f(\mathbf{x}_0, \mathbf{x}_1) \oplus g(\mathbf{x}_0, \mathbf{x}_1))}{\partial \mathbf{x}_0} = \frac{\partial f(\mathbf{x}_0, \mathbf{x}_1)}{\partial \mathbf{x}_0} \oplus \frac{\partial g(\mathbf{x}_0, \mathbf{x}_1)}{\partial \mathbf{x}_0} , \tag{2.38}$$

$$\min_{\mathbf{x}_0} (f(\mathbf{x}_0, \mathbf{x}_1) \wedge g(\mathbf{x}_0, \mathbf{x}_1)) = \min_{\mathbf{x}_0} f(\mathbf{x}_0, \mathbf{x}_1) \wedge \min_{\mathbf{x}_0} g(\mathbf{x}_0, \mathbf{x}_1) , \tag{2.39}$$

$$\max_{\mathbf{x}_0} (f(\mathbf{x}_0, \mathbf{x}_1) \vee g(\mathbf{x}_0, \mathbf{x}_1)) = \max_{\mathbf{x}_0} f(\mathbf{x}_0, \mathbf{x}_1) \vee \max_{\mathbf{x}_0} g(\mathbf{x}_0, \mathbf{x}_1) , \tag{2.40}$$

$$\frac{\partial}{\partial \mathbf{x}_0} \left(\frac{\partial f(\mathbf{x}_0, \mathbf{x}_1)}{\partial \mathbf{x}_0} \right) = 0 , \tag{2.41}$$

$$\min_{\mathbf{x}_0} \left(\min_{\mathbf{x}_0} f(\mathbf{x}_0, \mathbf{x}_1) \right) = \min_{\mathbf{x}_0} f(\mathbf{x}_0, \mathbf{x}_1) , \tag{2.42}$$

$$\max_{\mathbf{x}_0} \left(\max_{\mathbf{x}_0} f(\mathbf{x}_0, \mathbf{x}_1) \right) = \max_{\mathbf{x}_0} f(\mathbf{x}_0, \mathbf{x}_1) . \tag{2.43}$$

These relations follow directly from the definitions of the vectorial derivative operations and are valid for the corresponding simple derivative operations if the vector $\mathbf{x}_0$ is replaced by the variable x_i.

2.4   *m*-FOLD DERIVATIVE OPERATIONS

Now we consider a sequence of simple derivative operations. Since simple derivatives of $f(x_1, x_2, ..., x_n)$ with regard to x_i are again Boolean functions, further simple derivatives of the same type with regard to another variable can be calculated. It is known from (2.2), (2.4), and (2.6) that simple derivatives do not depend on the variable that already has been used for the derivation. Consequently, the result of m-fold differential operations of $f(x_1, x_2, ..., x_m, x_{m+1}, ..., x_n)$ with regard to $\mathbf{x}_0 = (x_1, ..., x_m)$ depends only on $n - m$ variables $\mathbf{x}_1 = (x_{m+1}, ..., x_n)$. Therefore m-fold derivative operations describe properties of whole subspaces specified by $\mathbf{x}_1 = const$.

Definition 2.14   Let $\mathbf{x}_0 = (x_1, x_2, ..., x_m)$, $\mathbf{x}_1 = (x_{m+1}, x_{m+2}, ..., x_n)$ be two disjoint sets of Boolean variables, and $f(\mathbf{x}_0, \mathbf{x}_1) = f(x_1, x_2, ..., x_n) = f(\mathbf{x})$ a Boolean function of n variables, then

$$\frac{\partial^m f(\mathbf{x}_0, \mathbf{x}_1)}{\partial x_1 \partial x_2 \ldots \partial x_m} = \frac{\partial}{\partial x_m} \left(\ldots \left(\frac{\partial}{\partial x_2} \left(\frac{\partial f(\mathbf{x}_0, \mathbf{x}_1)}{\partial x_1} \right) \right) \ldots \right) \tag{2.44}$$

is the *m-fold derivative* of the Boolean function $f(\mathbf{x}_0, \mathbf{x}_1)$ with regard to the subset of variables $\mathbf{x}_0$.

The m-fold derivative is a Boolean function that is equal to 1 for such subspaces $\mathbf{x}_1 = const$ where the function $f(\mathbf{x}_0, \mathbf{x}_1 = const)$ has an odd number of function values 1. The following example shows this property.

Example 2.15   Let $f(a, b, c, d) = b\,d \vee \overline{a}\,(c \vee \overline{d})$ be a given Boolean function. Based on Def. (2.44) and Th. (2.2), the 2-fold derivative $\frac{\partial^2 f(a,b,c,d)}{\partial c\,\partial d}$ can be calculated as follows:

$$\begin{aligned}
\frac{\partial^2 f(a, b, c, d)}{\partial c\,\partial d} &= \frac{\partial}{\partial d} \left(f(a, b, c = 0, d) \oplus f(a, b, c = 1, d) \right) \\
&= f(a, b, c = 0, d = 0) \oplus f(a, b, c = 1, d = 0)\oplus \\
&\quad f(a, b, c = 0, d = 1) \oplus f(a, b, c = 1, d = 1) \\
&= (\overline{a}) \oplus (\overline{a}) \oplus (b) \oplus (b \vee \overline{a}) \\
&= b \oplus b \oplus \overline{a} \oplus \overline{a}\,b \\
&= \overline{a}\,(1 \oplus b) \\
&= \overline{a}\,\overline{b}\,.
\end{aligned}$$

In order to evaluate the properties of the 2-fold derivative, we calculate the four subfunctions of $f(a, b, c, d)$ depending on (c, d),

$$f(a = 0, b = 0, c, d) = c \vee \overline{d}\,, \tag{2.45}$$

$$f(a = 0, b = 1, c, d) = d \vee c \vee \overline{d} = 1\,, \tag{2.46}$$

$$f(a = 1, b = 0, c, d) = 0\,, \tag{2.47}$$

$$f(a = 1, b = 1, c, d) = d\,, \tag{2.48}$$

and we observe that only in (2.45) three function values 1 exist. Because this is the only situation with an odd number of function values 1, the 2-fold derivative is equal to 1 only for $(a = 0, b = 0)$.

Definition 2.16   Let $\mathbf{x}_0 = (x_1, x_2, ..., x_m)$, $\mathbf{x}_1 = (x_{m+1}, x_{m+2}, ..., x_n)$ be two disjoint sets of Boolean variables, and $f(\mathbf{x}_0, \mathbf{x}_1) = f(x_1, x_2, ..., x_n) = f(\mathbf{x})$ a Boolean function of n variables, then

$$\min_{\mathbf{x}_0}{}^m f(\mathbf{x}_0, \mathbf{x}_1) = \min_{x_m} \left(... \left(\min_{x_2} \left(\min_{x_1} f(\mathbf{x}_0, \mathbf{x}_1) \right) \right) ... \right) \qquad (2.49)$$

is the *m-fold minimum* of the Boolean function $f(\mathbf{x}_0, \mathbf{x}_1)$ with regard to the subset of variables $\mathbf{x}_0$.

The m-fold minimum is a Boolean function that is equal to 1 for such subspaces $\mathbf{x}_1 = const$ where the function $f(\mathbf{x}_0, \mathbf{x}_1 = const)$ is constant equal to 1 for all patterns of the remaining variables $\mathbf{x}_0$.

Example 2.17   We take again the function $f(a, b, c, d) = b\,d \vee \bar{a}\,(c \vee \bar{d})$. Based on Def. (2.49) and Th. (2.4), the 2-fold minimum $\min^2_{(c,d)} f(a, b, c, d)$ can be calculated as follows:

$$\begin{aligned}
\min^2_{(c,d)} f(a, b, c, d) &= \min_d \left(f(a, b, c = 0, d) \wedge f(a, b, c = 1, d) \right) \\
&= f(a, b, c = 0, d = 0) \wedge f(a, b, c = 1, d = 0) \wedge \\
&\quad\ f(a, b, c = 0, d = 1) \wedge f(a, b, c = 1, d = 1) \\
&= (\bar{a}) \wedge (\bar{a}) \wedge (b) \wedge (b \vee \bar{a}) \\
&= \bar{a}\,b \vee \bar{a}\,b \\
&= \bar{a}\,b \,.
\end{aligned}$$

This result is confirmed by (2.46). Only the subfunction $f(a = 0, b = 1, c, d)$ is constant equal to 1.

It follows from (2.11) and Def. (2.49) that the m-fold minimum of a function with regard to a set of variables is less than an $(m - 1)$-fold minimum of the same function with regard to a subset of this set of variables:

$$\min_{(x_i, \mathbf{x}_0)}{}^m f(x_i, \mathbf{x}_0, \mathbf{x}_1) \leq \min_{\mathbf{x}_0}{}^{m-1} f(x_i, \mathbf{x}_0, \mathbf{x}_1) \leq f(x_i, \mathbf{x}_0, \mathbf{x}_1) \,. \qquad (2.50)$$

Definition 2.18   Let $\mathbf{x}_0 = (x_1, x_2, ..., x_m)$, $\mathbf{x}_1 = (x_{m+1}, x_{m+2}, ..., x_n)$ be two disjoint sets of Boolean variables and $f(\mathbf{x}_0, \mathbf{x}_1) = f(x_1, x_2, ..., x_n) = f(\mathbf{x})$ a Boolean function of n variables then

$$\max_{\mathbf{x}_0}{}^m f(\mathbf{x}_0, \mathbf{x}_1) = \max_{x_m} \left(... \left(\max_{x_2} \left(\max_{x_1} f(\mathbf{x}_0, \mathbf{x}_1) \right) \right) ... \right) \qquad (2.51)$$

is the *m-fold maximum* of the Boolean function $f(\mathbf{x}_0, \mathbf{x}_1)$ with regard to the subset of variables $\mathbf{x}_0$.

The *m*-fold maximum is a Boolean function that is equal to 1 for such subspaces $\mathbf{x}_1 = const$ where at least one function value of the function $f(\mathbf{x}_0, \mathbf{x}_1 = const)$ is equal to 1.

Example 2.19 We take again the same function $f(a, b, c, d) = b\,d \vee \overline{a}\,(c \vee \overline{d})$ and calculate the 2-fold maximum $\max^2_{(c,d)} f(a, b, c, d)$ based on Def. (2.51) and Th. (2.6):

$$
\begin{aligned}
\max^2_{(c,d)} f(a, b, c, d) &= \max_d \left(f(a, b, c = 0, d) \vee f(a, b, c = 1, d) \right) \\
&= f(a, b, c = 0, d = 0) \vee f(a, b, c = 1, d = 0) \vee \\
&\quad\; f(a, b, c = 0, d = 1) \vee f(a, b, c = 1, d = 1) \\
&= (\overline{a}) \vee (\overline{a}) \vee (b) \vee (b \vee \overline{a}) \\
&= \overline{a} \vee b \, .
\end{aligned}
$$

This result is confirmed by (2.45), (2.46), and (2.48). In these three subspaces exist function values 1 of $f(a, b, c, d)$. As shown in (2.47), $f(a = 1, b = 0, c, d)$ is constant equal to 0 and, thus, the conjunction $a \wedge \overline{b}$ does not appear in the result of the calculated *m*-fold maximum.

It follows from (2.11) and Def. (2.51) that the *m*-fold maximum of a function with regard to a set of variables is larger than an $(m - 1)$-fold maximum of the same function with regard to a subset of variables:

$$
f(x_i, \mathbf{x}_0, \mathbf{x}_1) \leq \max_{\mathbf{x}_0}{}^{m-1} f(x_i, \mathbf{x}_0, \mathbf{x}_1) \leq \max_{(x_i, \mathbf{x}_0)}{}^{m} f(x_i, \mathbf{x}_0, \mathbf{x}_1) \, . \tag{2.52}
$$

Analogously to Eqs. (2.7) and (2.27), another *m*-fold derivative operation can be defined.

Definition 2.20 Let $\mathbf{x}_0 = (x_1, x_2, ..., x_m)$, $\mathbf{x}_1 = (x_{m+1}, x_{m+2}, ..., x_n)$ be two disjoint sets of Boolean variables and $f(\mathbf{x}_0, \mathbf{x}_1) = f(x_1, x_2, ..., x_n) = f(\mathbf{x})$ a Boolean function of *n* variables, then

$$
\Delta_{\mathbf{x}_0} f(\mathbf{x}_0, \mathbf{x}_1) = \min_{\mathbf{x}_0}{}^m f(\mathbf{x}_0, \mathbf{x}_1) \oplus \max_{\mathbf{x}_0}{}^m f(\mathbf{x}_0, \mathbf{x}_1) \tag{2.53}
$$

is the Δ - operation of the Boolean function $f(\mathbf{x}_0, \mathbf{x}_1)$ with regard to the set of variables $\mathbf{x}_0$.

It follows from (2.53), (2.50), and (2.52) that $\Delta_{\mathbf{x}_0} f(\mathbf{x}_0, \mathbf{x}_1)$ is equal to 1 if both

$$
\min_{\mathbf{x}_0}{}^m f(\mathbf{x}_0, \mathbf{x}_1) = 0 \quad \text{and} \quad \max_{\mathbf{x}_0}{}^m f(\mathbf{x}_0, \mathbf{x}_1) = 1 \, .
$$

For that reason the Δ - operation characterizes such subspaces where the function is not constant.

Example 2.21   We reuse the results from the previous examples and calculate the Δ - operation of $f(a, b, c, d) = b\,d \vee \bar{a}\,(c \vee \bar{d})$ with regard to (c, d):

$$\Delta_{(c,d)} f(a, b, c, d) = \min^2_{(c,d)} f(a, b, c, d) \oplus \max^2_{(c,d)} f(a, b, c, d)$$
$$= (\bar{a}\,b) \oplus (\bar{a} \vee b)$$
$$= \bar{a}\,b \oplus \bar{a} \oplus b \oplus \bar{a}\,b$$
$$= \bar{a} \oplus b\,.$$
$$= a \odot b\,.$$

This result is confirmed by (2.45) and (2.48). The subfunctions $f(a = 0, b = 0, c, d)$ and $f(a = 1, b = 1, c, d)$ are not constant.

There is an important relation between the Δ - operation and the vectorial derivative operations.

Theorem 2.22   *Let* $\mathbf{x}_0 = (x_1, x_2, ..., x_m)$, $\mathbf{x}_1 = (x_{m+1}, x_{m+2}, ..., x_n)$ *be two disjoint sets of Boolean variables,* $P(\mathbf{x}_0)$ *the power set of* $\mathbf{x}_0$, *and* $f(\mathbf{x}_0, \mathbf{x}_1) = f(x_1, x_2, ..., x_n) = f(\mathbf{x})$ *a Boolean function of* n *variables, then it holds that*

$$\Delta_{\mathbf{x}_0} f(\mathbf{x}_0, \mathbf{x}_1) = \bigvee_{\mathbf{y} \in P(\mathbf{x}_0)} \frac{\partial f(\mathbf{x}_0, \mathbf{x}_1)}{\partial \mathbf{y}}\,. \tag{2.54}$$

Proof. Using (2.20) Eq. (2.54) can be transformed into (2.55):

$$\Delta_{\mathbf{x}_0} f(\mathbf{x}_0, \mathbf{x}_1) = \bigvee_{\mathbf{c} \in B^m} (f(\mathbf{x}_0, \mathbf{x}_1) \oplus f(\mathbf{x}_0 \oplus \mathbf{c}, \mathbf{x}_1))\,. \tag{2.55}$$

If $f(\mathbf{x}_0, \mathbf{x}_1 = \mathbf{c}_1)$ is constant equal to 0 then $f(\mathbf{x}_0 \oplus \mathbf{c}, \mathbf{x}_1 = \mathbf{c}_1)$ is constant equal to 0 too. Thus, $\Delta_{\mathbf{x}_0} f(\mathbf{x}_0, \mathbf{x}_1 = \mathbf{c}_1) = 0$ in (2.55) and equal to the definition (2.53).

If $f(\mathbf{x}_0, \mathbf{x}_1 = \mathbf{c}_1)$ is constant equal to 1, then $f(\mathbf{x}_0 \oplus \mathbf{c}, \mathbf{x}_1 = \mathbf{c}_1)$ is also constant equal to 1. Thus, $\Delta_{\mathbf{x}_0} f(\mathbf{x}_0, \mathbf{x}_1 = \mathbf{c}_1) = 0$ in (2.55) and both $\min^m_{\mathbf{x}_0} f(\mathbf{x}_0, \mathbf{x}_1 = \mathbf{c}_1)$ and $\max^m_{\mathbf{x}_0} f(\mathbf{x}_0, \mathbf{x}_1 = \mathbf{c}_1)$ are equal to 1 so that $\Delta_{\mathbf{x}_0} f(\mathbf{x}_0, \mathbf{x}_1 = \mathbf{c}_1)$ of (2.53) is equal to 0 too.

Finally, we assume that $f(\mathbf{x}_0, \mathbf{x}_1 = \mathbf{c}_1)$ is not constant. In this case $f(\mathbf{x}_0, \mathbf{x}_1 = \mathbf{c}_1)$ differs from $f(\mathbf{x}_0 \oplus \mathbf{c}, \mathbf{x}_1 = \mathbf{c}_1)$ at least once so that $\Delta_{\mathbf{x}_0} f(\mathbf{x}_0, \mathbf{x}_1 = \mathbf{c}_1)$ in (2.55) becomes equal to 1. The same function value occurs in (2.53) because under this condition $\min^m_{\mathbf{x}_0} f(\mathbf{x}_0, \mathbf{x}_1 = \mathbf{c}_1) = 0$ and $\max^m_{\mathbf{x}_0} f(\mathbf{x}_0, \mathbf{x}_1 = \mathbf{c}_1) = 1$. This proves the theorem. $\square$

Fig. 2.3 summarizes the results of all *m*-fold derivative operations of $f(a, b, c, d)$ with regard to (c, d), calculated in the examples above. Based on Defs. (2.49), (2.44), and (2.51), the *m*-fold derivative operations are calculated iteratively. The small arrows in the Karnaugh-map of $f(a, b, c, d)$ on top of Fig. 2.3 indicate the pairs of function values which are connected by an $\wedge$ -, $\oplus$ - or $\vee$ - operation, in order to get the simple minimum, simple derivative, or simple maximum with regard to the variable d as visualized in the second row of Fig. 2.3. The result of *m*-fold derivative operations does not depend on the order that has been used to calculate the simple derivative operations. For an easy visualization we start with the variable d.

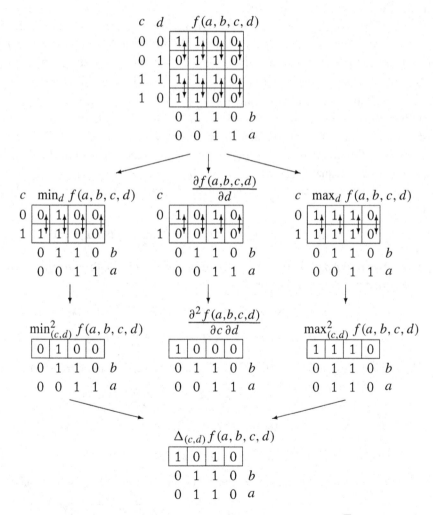

Figure 2.3: The Karnaugh-maps of the function $f(a, b, c, d) = b d \vee \overline{a} (c \vee \overline{d})$ and all 2-fold derivative operations with regard to (c, d).

In the second step, we calculate the three different simple derivative operations with regard to the variable c in order to get the m-fold derivative operations with regard to (c, d) as shown in the third row of Fig. 2.3. The small arrows in the Karnaugh-maps of the second row of Fig. 2.3 indicate the pairs of function values which have to be compared in order to get the results of the required m-fold derivative operations.

Finally, the Δ - operation is calculated by an $\oplus$ - operation between corresponding values of the first and the last Karnaugh-map in the third row of Fig. 2.3. This is a direct application of definition (2.53).

The 2-fold derivative operations with regard to (c, d) evaluate subspaces visualized by the columns of the Karnaugh-map of $f(a, b, c, d)$ in Fig. 2.3. The properties of the m-fold derivative operations can be recognized if their Karnaugh-maps are compared with the Karnaugh-map of $f(a, b, c, d)$. All function values in the second column of the Karnaugh-map of $f(a, b, c, d)$ are equal to 1. Therefore the function value 1 appears for $\min^2_{(c,d)} f(a = 0, b = 1, c, d)$. The only column of the Karnaugh-map of $f(a, b, c, d)$ which includes an odd number of function values 1 is the first one. Thus, the 2-fold derivative $\frac{\partial^2 f(a,b,c,d)}{\partial c \partial d}$ is equal to 1 for $(a = 0, b = 0)$. The values 1 in the first three columns of the Karnaugh-map of $\max^2_{(c,d)} f(a, b, c, d)$ indicate that there is at least one function value 1 in the associated subspaces of $f(a, b, c, d)$. Both function values, 0 and 1, appear in the first and third column of the Karnaugh-map of $f(a, b, c, d)$. For that reason $\Delta_{(c,d)} f(a, b, c, d)$ is equal to 1 for $(a = 0, b = 0)$ and $(a = 1, b = 1)$.

There are many relations between m-fold derivative operations. The general Eq. (2.56) follows from Def. (2.53):

$$\min_{\mathbf{x}_0}^m f(\mathbf{x}_0, \mathbf{x}_1) \oplus \Delta_{\mathbf{x}_0} f(\mathbf{x}_0, \mathbf{x}_1) \oplus \max_{\mathbf{x}_0}^m f(\mathbf{x}_0, \mathbf{x}_1) = 0 . \tag{2.56}$$

Using two of the m-fold derivative operations of (2.56), the third m-fold derivative operation can be calculated by (2.53), (2.57), or (2.58):

$$\min_{\mathbf{x}_0}^m f(\mathbf{x}_0, \mathbf{x}_1) = \Delta_{\mathbf{x}_0} f(\mathbf{x}_0, \mathbf{x}_1) \oplus \max_{\mathbf{x}_0}^m f(\mathbf{x}_0, \mathbf{x}_1) , \tag{2.57}$$

$$\max_{\mathbf{x}_0}^m f(\mathbf{x}_0, \mathbf{x}_1) = \Delta_{\mathbf{x}_0} f(\mathbf{x}_0, \mathbf{x}_1) \oplus \min_{\mathbf{x}_0}^m f(\mathbf{x}_0, \mathbf{x}_1) . \tag{2.58}$$

Eq. (2.59) follows from (2.50) and (2.52). Relation (2.59) can be transformed into Eq. (2.60):

$$\min_{\mathbf{x}_0}^m f(\mathbf{x}_0, \mathbf{x}_1) \leq f(\mathbf{x}_0, \mathbf{x}_1) \leq \max_{\mathbf{x}_0}^m f(\mathbf{x}_0, \mathbf{x}_1) , \tag{2.59}$$

$$\min_{\mathbf{x}_0}^m f(\mathbf{x}_0, \mathbf{x}_1) \wedge \overline{\max_{\mathbf{x}_0}^m f(\mathbf{x}_0, \mathbf{x}_1)} = 0 . \tag{2.60}$$

As shown below, (2.61) follows from Def. (2.53) and (2.60):

$$\Delta_{\mathbf{x}_0} f(\mathbf{x}_0, \mathbf{x}_1) \wedge \overline{\max_{\mathbf{x}_0}^m f(\mathbf{x}_0, \mathbf{x}_1)} = (\min_{\mathbf{x}_0}^m f(\mathbf{x}_0, \mathbf{x}_1) \oplus \max_{\mathbf{x}_0}^m f(\mathbf{x}_0, \mathbf{x}_1)) \wedge \overline{\max_{\mathbf{x}_0}^m f(\mathbf{x}_0, \mathbf{x}_1)}$$

$$= \min_{\mathbf{x}_0}^m f(\mathbf{x}_0, \mathbf{x}_1) \wedge \overline{\max_{\mathbf{x}_0}^m f(\mathbf{x}_0, \mathbf{x}_1)}$$

$$= 0 . \tag{2.61}$$

Furthermore, the orthogonality (2.62) between the m-fold minimum and the Δ - operation follows from definition (2.53) and rule (2.60):

$$
\begin{aligned}
\min_{\mathbf{x}_0}^m f(\mathbf{x}_0, \mathbf{x}_1) \wedge \Delta_{\mathbf{x}_0} f(\mathbf{x}_0, \mathbf{x}_1) &= \min_{\mathbf{x}_0}^m f(\mathbf{x}_0, \mathbf{x}_1) \wedge (\min_{\mathbf{x}_0}^m f(\mathbf{x}_0, \mathbf{x}_1) \oplus \max_{\mathbf{x}_0}^m f(\mathbf{x}_0, \mathbf{x}_1)) \\
&= \min_{\mathbf{x}_0}^m f(\mathbf{x}_0, \mathbf{x}_1) \oplus (\min_{\mathbf{x}_0}^m f(\mathbf{x}_0, \mathbf{x}_1) \wedge \max_{\mathbf{x}_0}^m f(\mathbf{x}_0, \mathbf{x}_1)) \\
&= \min_{\mathbf{x}_0}^m f(\mathbf{x}_0, \mathbf{x}_1) \wedge (1 \oplus \max_{\mathbf{x}_0}^m f(\mathbf{x}_0, \mathbf{x}_1)) \\
&= \min_{\mathbf{x}_0}^m f(\mathbf{x}_0, \mathbf{x}_1) \wedge \overline{\max_{\mathbf{x}_0}^m f(\mathbf{x}_0, \mathbf{x}_1)} \\
&= 0 .
\end{aligned}
\tag{2.62}
$$

Alternative rules to calculate m-fold derivative operations can be based on (2.53), (2.57), and (2.58), and take advantage from (2.60), (2.61), and (2.62):

$$
\Delta_{\mathbf{x}_0} f(\mathbf{x}_0, \mathbf{x}_1) = \max_{\mathbf{x}_0}^m f(\mathbf{x}_0, \mathbf{x}_1) \wedge \overline{\min_{\mathbf{x}_0}^m f(\mathbf{x}_0, \mathbf{x}_1)} ,
\tag{2.63}
$$

$$
\min_{\mathbf{x}_0}^m f(\mathbf{x}_0, \mathbf{x}_1) = \max_{\mathbf{x}_0}^m f(\mathbf{x}_0, \mathbf{x}_1) \wedge \overline{\Delta_{\mathbf{x}_0} f(\mathbf{x}_0, \mathbf{x}_1)} ,
\tag{2.64}
$$

$$
\max_{\mathbf{x}_0}^m f(\mathbf{x}_0, \mathbf{x}_1) = \min_{\mathbf{x}_0}^m f(\mathbf{x}_0, \mathbf{x}_1) \vee \Delta_{\mathbf{x}_0} f(\mathbf{x}_0, \mathbf{x}_1) .
\tag{2.65}
$$

Further properties of m-fold derivative operations are listed below. These formulas show negated functions, several functions and repeated applications of the same operation:

$$
\frac{\partial^m f(\mathbf{x}_0, \mathbf{x}_1)}{\partial x_1 \partial x_2 ... \partial x_m} = \frac{\partial^m \overline{f(\mathbf{x}_0, \mathbf{x}_1)}}{\partial x_1 \partial x_2 ... \partial x_m} ,
\tag{2.66}
$$

$$
\Delta_{\mathbf{x}_0} f(\mathbf{x}_0, \mathbf{x}_1) = \Delta_{\mathbf{x}_0} \overline{f(\mathbf{x}_0, \mathbf{x}_1)} ,
\tag{2.67}
$$

$$
\min_{\mathbf{x}_0}^m f(\mathbf{x}_0, \mathbf{x}_1) = \overline{\max_{\mathbf{x}_0}^m \overline{f(\mathbf{x}_0, \mathbf{x}_1)}} ,
\tag{2.68}
$$

$$
\max_{\mathbf{x}_0}^m f(\mathbf{x}_0, \mathbf{x}_1) = \overline{\min_{\mathbf{x}_0}^m \overline{f(\mathbf{x}_0, \mathbf{x}_1)}} ,
\tag{2.69}
$$

$$
\frac{\partial^m (f(\mathbf{x}) \oplus g(\mathbf{x}))}{\partial x_1 \partial x_2 ... \partial x_m} = \frac{\partial^m f(\mathbf{x})}{\partial x_1 \partial x_2 ... \partial x_m} \oplus \frac{\partial^m g(\mathbf{x})}{\partial x_1 \partial x_2 ... \partial x_m} ,
\tag{2.70}
$$

$$
\min_{\mathbf{x}_0}^m (f(\mathbf{x}) \wedge g(\mathbf{x})) = \min_{\mathbf{x}_0}^m f(\mathbf{x}) \wedge \min_{\mathbf{x}_0}^m g(\mathbf{x}) ,
\tag{2.71}
$$

$$
\max_{\mathbf{x}_0}^m (f(\mathbf{x}) \vee g(\mathbf{x})) = \max_{\mathbf{x}_0}^m f(\mathbf{x}) \vee \max_{\mathbf{x}_0}^m g(\mathbf{x}) ,
\tag{2.72}
$$

$$
\Delta_{\mathbf{x}_0} \left(\Delta_{\mathbf{x}_0} f(\mathbf{x}_0, \mathbf{x}_1) \right) = 0,
\tag{2.73}
$$

$$
\min_{\mathbf{x}_0}^m \left(\min_{\mathbf{x}_0}^m f(\mathbf{x}) \right) = \min_{\mathbf{x}_0}^m f(\mathbf{x}) ,
\tag{2.74}
$$

$$
\max_{\mathbf{x}_0}^m \left(\min_{\mathbf{x}_0}^m f(\mathbf{x}) \right) = \max_{\mathbf{x}_0}^m f(\mathbf{x}) .
\tag{2.75}
$$

Note. In the 0-fold derivative operations, the vector $\mathbf{x}_0$ is empty so that both the 0-fold Δ - operation and the 0-fold derivative becomes 0 and the function $f(\mathbf{x}_1)$ itself is the result of the 0-fold minimum or the 0-fold maximum.

SUMMARY

The Boolean Differential Calculus extends the Boolean Algebra such that *changes* of function values are evaluated. These changes can be focused to pairs of function values or whole subspaces. Boolean derivatives are calculated for given Boolean functions. Each Boolean derivative is again a Boolean function with special properties. While the direction of change is fixed for all kinds of derivative operations, several directions of change are taken into account for differential operations. Commonly the Boolean Algebra and the Boolean Differential Calculus are strong tools to solve the tasks around digital circuits and systems efficiently.

EXERCISES

2.1 The permitted behavior of a digital circuit is defined by the equation:

$$F(a, b, c, y) = b\bar{c}\,y \oplus (\bar{c}\,\bar{y} \vee a\,y \vee c\,y) = 1 \ . \tag{2.76}$$

A realizable circuit requires that Eq. (2.76) is solvable with regard to the variable $y = f(a, b, c)$. This condition becomes true if the simple maximum of $F(a, b, c, y)$ with regard to y is equal to 1. Calculate this maximum and verify whether this condition is satisfied. The behavior is uniquely defined when the simple derivative of $F(a, b, c, y)$ with regard to y is equal to 1. Calculate this derivative and verify whether $F(a, b, c, y)$ defines a unique function $y = f(a, b, c)$. If the given behavior describes the digital circuit not uniquely, the don't care function $\varphi(a, b, c)$ is equal to 1 for such patterns (a, b, c) where $f(a, b, c)$ can be chosen arbitrarily. The function $\varphi(a, b, c)$ is found as the result of the simple minimum $F(a, b, c, y)$ with regard to y. Calculate this minimum and determine how many different functions $y = f(a, b, c)$ satisfy the permitted behavior (2.76). Solve this task first by paper and pen and thereafter using XBOOLE.

2.2 Two functions $f_1(\mathbf{x})$ and $f_2(\mathbf{x})$ are dual to each other if $f_1(\mathbf{x}) = \overline{f_2(\overline{\mathbf{x}})}$. One function $f(\mathbf{x})$ is called self-dual if $f(\mathbf{x}) = \overline{f(\overline{\mathbf{x}})}$. Hence, the vectorial derivative with regard to all variables $\mathbf{x}$ of a self-dual function $f(\mathbf{x})$ is equal to 1. Calculate the vectorial derivative and verify whether the function (2.77) is self-dual.

$$f(a, b, c, d) = a\bar{b} \vee a\bar{c} \vee \bar{b}\bar{c}\bar{d} \vee a\bar{d} \tag{2.77}$$

2.3 Are there subspaces $(a, b) = const$ where the function (2.77) is constant? Solve this problem by means of an applicable 2-fold derivative operation.

2.4 A function $f(a, b, c, d)$ can be decomposed into $f(a, b, c, d) = g(a, b, c) \oplus h(b, c, d)$ if the 2-fold derivative of $f(a, b, c, d)$ with regard to a and d is equal to 0. Calculate the 2-fold derivative of function (2.77) with regard to a and d and verify whether this EXOR-decomposition is possible.

CHAPTER 3

Boolean Differential Equations

3.1 INTRODUCTION

A *Boolean Differential Equation* (BDE) is a Boolean equation that includes one or more functions and operations of the Boolean Differential Calculus that are applied to these functions.

When all the functions are known, then these operations can be applied such that only a Boolean equation remains without any functions. Such BDEs appear in practical applications, but require for their solution only the knowledge of the operations of the Boolean Differential Calculus.

New approaches are required to solve a BDE in the case that the functions within the equations are unknown. Such approaches are explained in this chapter starting with very simple BDEs that include a single derivative operation and extended step by step up to the most general BDEs.

3.2 BOOLEAN DIFFERENTIAL EQUATIONS OF A SINGLE SIMPLE DERIVATIVE

The reason of Boolean differential equations can be seen by a rather simple analysis. Each Boolean derivative is again a Boolean function. Hence, there is the Boolean differential equation

$$\frac{\partial f(a, b)}{\partial a} = g(a, b) \ . \tag{3.1}$$

This BDE can be met in two different situations.

First, we assume that the function

$$f(a, b) = a \vee b \tag{3.2}$$

is given. Using Def. (2.1) we get the simple derivative of the function (3.2) with regard to a as follows:

$$\begin{aligned}
\frac{\partial f(a, b)}{\partial a} &= (a \vee b) \oplus (\overline{a} \vee b) \\
&= (a \oplus b \oplus a b) \oplus (\overline{a} \oplus b \oplus \overline{a} b) \\
&= a \oplus \overline{a} \oplus b \oplus b \oplus a b \oplus \overline{a} b \\
&= 1 \oplus b(a \oplus \overline{a}) \\
&= \overline{b} \ .
\end{aligned} \tag{3.3}$$

Hence, the result of this simple derivative is the Boolean function

$$g(a, b) = \overline{b} \ . \tag{3.4}$$

The function $g(a, b)$ in (3.4) depends uniquely on the given function $f(a, b)$ in (3.2), the definition of the simple derivative (2.1) and the direction of change described by the taken variable a.

As a second possibility for the BDE (3.1) the function $g(a, b)$ (3.4) can be given. Because all calculation steps of (3.3) can also be executed in the reverse direction, the function $f(a, b)$ (3.2) is a solution of the Boolean differential equation (3.1) where the function $g(a, b)$ is defined by (3.4). Now the question arises whether the function $f(a, b)$ is uniquely defined by the function $g(a, b)$ (3.4) and the BDE (3.1). Due to the finite number of all Boolean functions and the chosen restriction to two variables a and b we can answer this question by the calculation of the simple derivatives of all Boolean function $f(a, b)$ with regard to a. Tab. 3.1 on page 51 shows this complete enumeration.

In the last column we see that the function $g(a, b) = \bar{b}$ is the result of the simple derivative of four different functions with regard to a. Hence, the function $f(a, b)$ can be chosen from the set of functions:

$$\left\{ a\bar{b}, a \vee b, \bar{a}\bar{b}, \bar{a} \vee b \right\} .$$

This set includes the function $f(a, b) = a \vee b$ initially used to calculate $g(a, b)$ as simple derivative with regard to a. Due to the other functions of the function set we learn from this simple example:

> **The solution of a Boolean differential equation is a**
>
> *set of Boolean functions.*

From Tab. 3.1 we get a second conclusion. Not all possible functions $g(a, b)$ appear as the result of a derivative of $f(a, b)$ with regard to a. As can be seen in this table, the function $g(a, b) = a b$, for instance, does not appear as the result of the simple derivative of any function $f(a, b)$ with regard to a. More precisely, we detect that all functions $g(a, b)$ which are the result of the simple derivative of any $f(a, b)$ are independent on the variable a, and consequently

$$\frac{\partial g(a, b)}{\partial a} = 0 \tag{3.5}$$

is the *integrability condition* for the BDE (3.1).

From Tab. 3.1 we get a third conclusion. Each function $g(a, b)$ appears exactly four times as the result of the simple derivative in the fourth column. This gives us a hint for a general formula which describes how the set of all solution functions of the BDE (3.1) can be calculated. As a special case of the BDE (3.1) the function $g(a, b) = 0$ can be chosen. A BDE with the value 0 on the right side is called *homogeneous restrictive* BDE. The solution functions of the homogeneous restrictive BDE (3.1) where $g(a, b) = 0$ can be taken from the second column of Tab. 3.1 as follows:

$$\left\{ 0, b, \bar{b}, 1 \right\} . \tag{3.6}$$

Using this set of functions $h(b)$ we find each solution function $f(a, b)$ of the general BDE (3.1) by

$$f(a, b) = a \wedge g(a, b) \oplus h(b) , \tag{3.7}$$

Table 3.1: Simple derivatives of all Boolean functions $f(a, b)$ with regard to a

$f(a, b)$		$\frac{\partial f(a,b)}{\partial a}$	
function vector	expression	function vector	expression
(0000)	0	(0000)	0
(0001)	$a \wedge b$	(0101)	b
(0010)	$a \wedge \overline{b}$	(1010)	$\overline{b}$
(0011)	a	(1111)	1
(0100)	$\overline{a} \wedge b$	(0101)	b
(0101)	b	(0000)	0
(0110)	$a \oplus b$	(1111)	1
(0111)	$a \vee b$	(1010)	$\overline{b}$
(1000)	$\overline{a} \wedge \overline{b}$	(1010)	$\overline{b}$
(1001)	$a \odot b$	(1111)	1
(1010)	$\overline{b}$	(0000)	0
(1011)	$a \vee \overline{b}$	(0101)	b
(1100)	$\overline{a}$	(1111)	1
(1101)	$\overline{a} \vee b$	(1010)	$\overline{b}$
(1110)	$\overline{a} \vee \overline{b}$	(0101)	b
(1111)	1	(0000)	0

if $g(a, b)$ holds the integrability condition (3.5) and $h(b)$ is any function of the solution set (3.6) of the associated homogeneous restrictive BDE of (3.1). The first term in (3.7) is a special solution of the general BDE (3.1), and the second term is the general solution of the associated homogeneous restrictive BDE.

For a detailed analysis we restricted the considerations to functions of two variables in the BDE (3.1). All the results can be generalized as follows: A BDE

$$\frac{\partial f(\mathbf{x})}{\partial x_i} = g(\mathbf{x}) , \tag{3.8}$$

with $\mathbf{x} = (x_i, \mathbf{x}_1)$ has the solution

$$f(\mathbf{x}) = x_i \wedge g(\mathbf{x}) \oplus h(\mathbf{x}) , \tag{3.9}$$

if $g(\mathbf{x})$ holds the integrability condition

$$\frac{\partial g(\mathbf{x})}{\partial x_i} = 0 , \tag{3.10}$$

and $h(\mathbf{x})$ are all solution functions of

$$\frac{\partial h(\mathbf{x})}{\partial x_i} = 0 . \tag{3.11}$$

The BDE (3.11) expresses that $h(\mathbf{x}) = h(x_i, \mathbf{x}_1)$ does not depend on x_i. Hence, each of the $2^{2^{|\mathbf{x}_1|}}$ functions $h(\mathbf{x}_1)$ can be used to construct a solution function by (3.9) for the BDE (3.8).

The set of functions $h(\mathbf{x})$ is defined by the homogeneous BDE (3.11) which expresses that $h(\mathbf{x})$ does not depend on the variable x_i. Therefore, the admissible set of functions $h(\mathbf{x})$ in (3.9) contains $2^{2^{|\mathbf{x}|-1}}$ functions. The conjecture that the decrease of the exponent by the value 1 is a special feature of the simple derivative in the BDE (3.8) is not true. As shown in the next subsection, a solvable BDE of a single vectorial derivative has the same number of solution functions.

3.3   BOOLEAN DIFFERENTIAL EQUATIONS OF A SINGLE VECTORIAL DERIVATIVE

The acquired knowledge for simple derivatives will now be extended to BDEs of a single *vectorial derivative*. The vectorial derivative is again a Boolean function. Hence, there is the Boolean differential equation:

$$\frac{\partial f(a, b)}{\partial(a, b)} = g(a, b) . \tag{3.12}$$

We follow the approach of the last subsection and solve this BDE regarding two different requirements.

First, we assume again that the function $f(a, b)$ (3.2) is given. Using Def. (2.20) we get the vectorial derivative of the function (3.2) with regard of the common change of a and b as follows:

$$
\begin{aligned}
\frac{\partial f(a, b)}{\partial (a, b)} &= (a \vee b) \oplus (\overline{a} \vee \overline{b}) \\
&= (a \oplus b \oplus a\, b) \oplus (\overline{a} \oplus \overline{b} \oplus \overline{a}\, \overline{b}) \\
&= a \oplus \overline{a} \oplus b \oplus \overline{b} \oplus a\, b \oplus \overline{a}\, \overline{b} \\
&= 1 \oplus 1 \oplus a\, b \oplus \overline{a}\, \overline{b} \\
&= a\, b \oplus \overline{a}\, \overline{b} \, .
\end{aligned}
\tag{3.13}
$$

The result of the vectorial derivative of $f(a, b)$ (3.2) with regard to (a, b) is the Boolean function

$$
g(a, b) = a \odot b \, .
\tag{3.14}
$$

The function $g(a, b)$ (3.14) of the vectorial derivative (3.13) is also uniquely defined. In contradiction to the simple derivative this vectorial derivative depends on both variables a and b.

Studying the reverse direction, we assume secondly that the function $g(a, b)$ (3.14) is given within the BDE (3.12). All calculation steps of (3.13) can be again executed in the reverse direction. Hence, the function $f(a, b)$ (3.2) is a solution of the Boolean differential equation (3.12) where the function $g(a, b)$ is defined by (3.14).

The function $g(a, b)$ (3.14) depends on both variables a and b. Hence the implicit restriction to $2^{2^1} = 4$ possible functions depending on a single variable as in the case of a simple derivative in the BDE (3.1) is not given. Tab. 3.2 on page 54 enumerates all vectorial derivatives in order to find out whether the function $f(a, b)$ (3.2) is uniquely defined by the BDE (3.12), or further solution functions of the BDE (3.12) for the given function (3.14) exist.

In the last column of Tab. 3.2 we see that the function $g(a, b) = a \odot b$ is the result of the vectorial derivative of four different functions $f(a, b)$ with regard to (a, b). Hence, the function $f(a, b)$ can be chosen from the set of functions:

$$
\left\{ a\, b, a \vee b, \overline{a}\, \overline{b}, \overline{a} \vee \overline{b} \right\} \, .
$$

For this set of functions we observe:

1. it includes the function $f(a, b) = a \vee b$ basically used to calculate $g(a, b)$ as vectorial derivative with regard to (a, b),

2. it is different to the set of functions of $f(a, b)$ as solution of the BDE (3.1) of a single simple derivative with regard to a, but

3. it also contains four solution functions as in the case of a single simple derivative within the BDE (3.1).

The solution of the BDE (3.12) of a single vectorial derivative confirms that generally the solution of a BDE is a set of Boolean functions. In Tab. 3.2 we observe that only four different

Table 3.2: Vectorial derivatives of all Boolean functions $f(a, b)$ with regard to (a, b)

$f(a, b)$		$\frac{\partial f(a,b)}{\partial (a,b)}$	
function vector	expression	function vector	expression
(0000)	0	(0000)	0
(0001)	$a \wedge b$	(1001)	$a \odot b$
(0010)	$a \wedge \overline{b}$	(0110)	$a \oplus b$
(0011)	a	(1111)	1
(0100)	$\overline{a} \wedge b$	(0110)	$a \oplus b$
(0101)	b	(1111)	1
(0110)	$a \oplus b$	(0000)	0
(0111)	$a \vee b$	(1001)	$a \odot b$
(1000)	$\overline{a} \wedge \overline{b}$	(1001)	$a \odot b$
(1001)	$a \odot b$	(0000)	0
(1010)	$\overline{b}$	(1111)	1
(1011)	$a \vee \overline{b}$	(0110)	$a \oplus b$
(1100)	$\overline{a}$	(1111)	1
(1101)	$\overline{a} \vee b$	(0110)	$a \oplus b$
(1110)	$\overline{a} \vee \overline{b}$	(1001)	$a \odot b$
(1111)	1	(0000)	0

functions $g(a, b) \in \{0, a \oplus b, a \odot b, 1\}$ appear as the result of the vectorial derivative of any function $f(a, b)$ with regard to (a, b). Hence, there must again be an *integrability condition* for the BDE (3.12) which is

$$\frac{\partial g(a, b)}{\partial (a, b)} = 0 . \tag{3.15}$$

A third observation of Tab. 3.2 is that each of the four functions $g(a, b)$ appears exactly four times as a result of a vectorial derivative of $f(a, b)$ with regard to (a, b). This property is consistent with the simple derivative because both the definition of the simple derivative (2.1) and the definition of the vectorial derivative (2.20) compare pairs of function values. Therefore, all solutions of the BDE (3.12) can be constructed in a similar way as in the case of the BDE (3.1) of a simple derivative.

As a special case of the BDE (3.12) the function $g(a, b) = 0$ can be chosen. This BDE with the value 0 on the right side is again called *homogeneous restrictive* BDE. The solution functions of the homogeneous restrictive BDE (3.12) with $g(a, b) = 0$ can be taken from the second column of Tab. 3.2 as follows:

$$\{0, a \oplus b, a \odot b, 1\} . \tag{3.16}$$

Using this set of functions $h(a, b)$, each solution function $f(a, b)$ of the general BDE (3.12) can be calculated by:

$$f(a, b) = a \wedge g(a, b) \oplus h(a, b) , \tag{3.17}$$

if $g(a, b)$ holds the integrability condition (3.15) and $h(a, b)$ is any function of the solution set (3.16) of the associated homogeneous restrictive BDE of (3.12). The first term in (3.17) is a special solution of the general BDE (3.12) and the second term is the general solution of the associated homogeneous restrictive BDE. The variable a in the first conjunction of (3.17) can be replaced by the other variable b of the vector (a, b) used to calculate the vectorial derivative in the BDE (3.12).

As in the case of the BDE of a single simple derivative the BDE of a single vectorial derivative can be generalized as follows: A BDE

$$\frac{\partial f(\mathbf{x})}{\partial \mathbf{x}_0} = g(\mathbf{x}) , \tag{3.18}$$

with $\mathbf{x} = (\mathbf{x}_0, \mathbf{x}_1), \mathbf{x}_0 = (x_1, x_2, \ldots, x_i, \ldots, x_k), \mathbf{x}_1 = (x_{k+1}, x_{k+2}, \ldots, x_n)$, has the solution

$$f(\mathbf{x}) = x_i \wedge g(\mathbf{x}) \oplus h(\mathbf{x}) , \tag{3.19}$$

if $g(\mathbf{x})$ holds the integrability condition

$$\frac{\partial g(\mathbf{x})}{\partial \mathbf{x}_0} = 0 , \tag{3.20}$$

and $h(\mathbf{x})$ are all solution functions of

$$\frac{\partial h(\mathbf{x})}{\partial \mathbf{x}_0} = 0 . \tag{3.21}$$

The homogeneous restrictive BDE (3.21) requires that only $|\mathbf{x}| - 1$ function values can be chosen arbitrarily. Therefore, the admissible set of solution functions $h(\mathbf{x})$ of (3.19) contains $2^{2^{|\mathbf{x}|-1}}$ functions.

3.4 BOOLEAN DIFFERENTIAL EQUATIONS RESTRICTED TO ALL VECTORIAL DERIVATIVES

3.4.1 ESSENCE OF BOOLEAN DIFFERENTIAL EQUATIONS

A more general Boolean differential equation contains more than one derivative of a Boolean function. We generalize the explored BDEs in this subsection such that the BDE can contain vectorial derivatives of one function and this function itself in expressions D_1 and D_2 on both sides of the BDE:

$$D_1\left(f(\mathbf{x}), \frac{\partial f(\mathbf{x})}{\partial x_1}, \frac{\partial f(\mathbf{x})}{\partial x_2}, \ldots, \frac{\partial f(\mathbf{x})}{\partial \mathbf{x}}\right) = D_2\left(f(\mathbf{x}), \frac{\partial f(\mathbf{x})}{\partial x_1}, \frac{\partial f(\mathbf{x})}{\partial x_2}, \ldots, \frac{\partial f(\mathbf{x})}{\partial \mathbf{x}}\right). \quad (3.22)$$

The included simple derivatives are considered as special vectorial derivatives. The restriction to BDEs with these derivatives allows us to conceive the essence of Boolean differential equations and facilitate the further generalization to BDEs with all other types of derivative operations.

The general BDE (3.22) can be transformed into a homogenous characteristic BDE:

$$D_1\left(f(\mathbf{x}), \frac{\partial f(\mathbf{x})}{\partial x_1}, \frac{\partial f(\mathbf{x})}{\partial x_2}, \ldots, \frac{\partial f(\mathbf{x})}{\partial \mathbf{x}}\right) \odot D_2\left(f(\mathbf{x}), \frac{\partial f(\mathbf{x})}{\partial x_1}, \frac{\partial f(\mathbf{x})}{\partial x_2}, \ldots, \frac{\partial f(\mathbf{x})}{\partial \mathbf{x}}\right) = 1$$

$$D_c\left(f(\mathbf{x}), \frac{\partial f(\mathbf{x})}{\partial x_1}, \frac{\partial f(\mathbf{x})}{\partial x_2}, \ldots, \frac{\partial f(\mathbf{x})}{\partial \mathbf{x}}\right) = 1, \quad (3.23)$$

or a homogenous restrictive BDE:

$$D_1\left(f(\mathbf{x}), \frac{\partial f(\mathbf{x})}{\partial x_1}, \frac{\partial f(\mathbf{x})}{\partial x_2}, \ldots, \frac{\partial f(\mathbf{x})}{\partial \mathbf{x}}\right) \oplus D_2\left(f(\mathbf{x}), \frac{\partial f(\mathbf{x})}{\partial x_1}, \frac{\partial f(\mathbf{x})}{\partial x_2}, \ldots, \frac{\partial f(\mathbf{x})}{\partial \mathbf{x}}\right) = 0$$

$$D_r\left(f(\mathbf{x}), \frac{\partial f(\mathbf{x})}{\partial x_1}, \frac{\partial f(\mathbf{x})}{\partial x_2}, \ldots, \frac{\partial f(\mathbf{x})}{\partial \mathbf{x}}\right) = 0. \quad (3.24)$$

In order to support the understanding of the essence of Boolean differential equations we explore in this subsection a simple BDE step by step in detail. For that reason, we use a method that emphasizes the feature of the BDE. This benefit must be paid by a time-consuming procedure. Using the detected knowledge an efficient method to solve such BDEs will be presented in subsection 3.4.3.

As example we use the Boolean differential equation (3.25). This BDE contains simple derivatives with regard to the variables a and b of the function $f(a, b)$ and the vectorial derivative of the same function with regard to both variables. It does not matter that the function $f(a, b)$ does not appear in the BDE. An equivalent transformation later on allows the introduction of this function into the BDE.

$$\frac{\partial f(a, b)}{\partial a} \vee \frac{\partial f(a, b)}{\partial b} = \overline{\frac{\partial f(a, b)}{\partial(a, b)}} \quad (3.25)$$

Equivalent transformations as known from Boolean equations can be applied to BDEs too. As a first step the given BDE (3.25) must be transformed into a homogeneous characteristic BDE.

This can be reached using an EXOR with the vectorial derivative of $f(a, b)$ with regard to (a, b).

$$\left(\frac{\partial f(a,b)}{\partial a} \vee \frac{\partial f(a,b)}{\partial b}\right) \oplus \frac{\partial f(a,b)}{\partial(a,b)} = \frac{\overline{\partial f(a,b)}}{\partial(a,b)} \oplus \frac{\partial f(a,b)}{\partial(a,b)}$$

$$\left(\frac{\partial f(a,b)}{\partial a} \vee \frac{\partial f(a,b)}{\partial b}\right) \oplus \frac{\partial f(a,b)}{\partial(a,b)} = 1 \oplus \frac{\partial f(a,b)}{\partial(a,b)} \oplus \frac{\partial f(a,b)}{\partial(a,b)}$$

$$\left(\frac{\partial f(a,b)}{\partial a} \vee \frac{\partial f(a,b)}{\partial b}\right) \oplus \frac{\partial f(a,b)}{\partial(a,b)} = 1 \tag{3.26}$$

In the second step the expression on the left-hand side of the homogeneous characteristic BDE (3.26) must be transformed into a disjunctive normal form of the existing derivatives. This can be done using identities of expressions.

$$\left(\frac{\partial f(a,b)}{\partial a} \vee \frac{\partial f(a,b)}{\partial b}\right) \oplus \frac{\partial f(a,b)}{\partial(a,b)} = 1$$

$$\overline{\left(\frac{\partial f(a,b)}{\partial a} \vee \frac{\partial f(a,b)}{\partial b}\right)} \wedge \frac{\partial f(a,b)}{\partial(a,b)} \vee \left(\frac{\partial f(a,b)}{\partial a} \vee \frac{\partial f(a,b)}{\partial b}\right) \wedge \frac{\overline{\partial f(a,b)}}{\partial(a,b)} = 1$$

$$\frac{\overline{\partial f(a,b)}}{\partial a} \frac{\overline{\partial f(a,b)}}{\partial b} \frac{\partial f(a,b)}{\partial(a,b)} \vee \frac{\partial f(a,b)}{\partial a} \frac{\overline{\partial f(a,b)}}{\partial(a,b)} \vee \frac{\partial f(a,b)}{\partial b} \frac{\overline{\partial f(a,b)}}{\partial(a,b)} = 1 \tag{3.27}$$

The first conjunction of (3.27) already contains all three derivatives as necessary for the BDE in normal form. The second conjunction must be extended by the disjunction of the simple derivative of $f(a, b)$ with regard to b and its complement. Similarly, the last conjunction must be extended with the disjunction of the simple derivative of $f(a, b)$ with regard to a and its complement. The application of the distributive law generates two conjunctions of all three derivatives. One of these conjunctions appears twice and can be removed once from the BDE without changing the solution due to the rule $x \vee x = x$ of the Boolean Algebra.

$$\frac{\overline{\partial f(a,b)}}{\partial a} \frac{\overline{\partial f(a,b)}}{\partial b} \frac{\partial f(a,b)}{\partial(a,b)} \vee \frac{\partial f(a,b)}{\partial a} \frac{\overline{\partial f(a,b)}}{\partial(a,b)} \vee \frac{\partial f(a,b)}{\partial b} \frac{\overline{\partial f(a,b)}}{\partial(a,b)} = 1$$

$$\frac{\partial f(a,b)}{\partial a} \left(\frac{\partial f(a,b)}{\partial b} \vee \frac{\overline{\partial f(a,b)}}{\partial b}\right) \frac{\overline{\partial f(a,b)}}{\partial(a,b)} \vee$$

$$\left(\frac{\partial f(a,b)}{\partial a} \vee \frac{\overline{\partial f(a,b)}}{\partial a}\right) \frac{\partial f(a,b)}{\partial b} \frac{\overline{\partial f(a,b)}}{\partial(a,b)} = 1$$

$$\frac{\overline{\partial f(a,b)}}{\partial a} \frac{\overline{\partial f(a,b)}}{\partial b} \frac{\partial f(a,b)}{\partial(a,b)} \vee$$

$$\frac{\partial f(a,b)}{\partial a} \frac{\partial f(a,b)}{\partial b} \frac{\overline{\partial f(a,b)}}{\partial(a,b)} \vee \frac{\partial f(a,b)}{\partial a} \frac{\overline{\partial f(a,b)}}{\partial b} \frac{\overline{\partial f(a,b)}}{\partial(a,b)} \vee$$

$$\frac{\partial f(a,b)}{\partial a} \frac{\partial f(a,b)}{\partial b} \frac{\overline{\partial f(a,b)}}{\partial(a,b)} \vee \frac{\overline{\partial f(a,b)}}{\partial a} \frac{\partial f(a,b)}{\partial b} \frac{\overline{\partial f(a,b)}}{\partial(a,b)} = 1$$

$$\frac{\overline{\partial f(a,b)}}{\partial a} \frac{\overline{\partial f(a,b)}}{\partial b} \frac{\partial f(a,b)}{\partial(a,b)} \vee \frac{\partial f(a,b)}{\partial a} \frac{\overline{\partial f(a,b)}}{\partial b} \frac{\partial f(a,b)}{\partial(a,b)} \vee$$

$$\frac{\partial f(a,b)}{\partial a} \frac{\partial f(a,b)}{\partial b} \frac{\overline{\partial f(a,b)}}{\partial(a,b)} \vee \frac{\overline{\partial f(a,b)}}{\partial a} \frac{\partial f(a,b)}{\partial b} \frac{\overline{\partial f(a,b)}}{\partial(a,b)} = 1 \qquad (3.28)$$

Due to the basic assumption the function itself is expected in the BDE. When the function of the derivatives does not appear in the BDE the normal form of the BDE can be extended by a disjunction of the function and its complement without influence on the solution.

$$\left(f(a,b) \vee \overline{f(a,b)}\right) \wedge \left(\frac{\overline{\partial f(a,b)}}{\partial a} \frac{\overline{\partial f(a,b)}}{\partial b} \frac{\partial f(a,b)}{\partial(a,b)} \vee \frac{\partial f(a,b)}{\partial a} \frac{\overline{\partial f(a,b)}}{\partial b} \frac{\overline{\partial f(a,b)}}{\partial(a,b)} \vee \right.$$

$$\left. \frac{\partial f(a,b)}{\partial a} \frac{\partial f(a,b)}{\partial b} \frac{\overline{\partial f(a,b)}}{\partial(a,b)} \vee \frac{\overline{\partial f(a,b)}}{\partial a} \frac{\partial f(a,b)}{\partial b} \frac{\overline{\partial f(a,b)}}{\partial(a,b)} \right) = 1$$

$$(3.29)$$

The application of the distributive law to the extended form (3.29) of the BDE generates the complete disjunctive normal form (3.30). Such a normal form can be constructed for each BDE in which only simple and vectorial derivatives of the same function and the function itself appear.

$$f(a,b) \frac{\overline{\partial f(a,b)}}{\partial a} \frac{\overline{\partial f(a,b)}}{\partial b} \frac{\partial f(a,b)}{\partial(a,b)} \vee f(a,b) \frac{\partial f(a,b)}{\partial a} \frac{\overline{\partial f(a,b)}}{\partial b} \frac{\overline{\partial f(a,b)}}{\partial(a,b)} \vee$$

$$f(a,b) \frac{\partial f(a,b)}{\partial a} \frac{\partial f(a,b)}{\partial b} \frac{\overline{\partial f(a,b)}}{\partial(a,b)} \vee f(a,b) \frac{\overline{\partial f(a,b)}}{\partial a} \frac{\partial f(a,b)}{\partial b} \frac{\overline{\partial f(a,b)}}{\partial(a,b)} \vee$$

$$\overline{f(a,b)} \frac{\overline{\partial f(a,b)}}{\partial a} \frac{\overline{\partial f(a,b)}}{\partial b} \frac{\partial f(a,b)}{\partial(a,b)} \vee \overline{f(a,b)} \frac{\partial f(a,b)}{\partial a} \frac{\overline{\partial f(a,b)}}{\partial b} \frac{\overline{\partial f(a,b)}}{\partial(a,b)} \vee$$

$$\overline{f(a,b)} \frac{\partial f(a,b)}{\partial a} \frac{\partial f(a,b)}{\partial b} \frac{\overline{\partial f(a,b)}}{\partial(a,b)} \vee \overline{f(a,b)} \frac{\overline{\partial f(a,b)}}{\partial a} \frac{\partial f(a,b)}{\partial b} \frac{\overline{\partial f(a,b)}}{\partial(a,b)} = 1 \qquad (3.30)$$

A solution function of the BDE (3.30) must satisfy this equation for each assignment of constant Boolean values to the variables a and b. Due to the finite number of elements of a Boolean space, each BDE can be evaluated for each assignment vector iteratively. In the case of the simple BDE (3.30) the assignments $(a,b) = (0,0)$, $(a,b) = (1,0)$, $(a,b) = (0,1)$, and $(a,b) = (1,1)$ must be applied.

Let us start with a detailed evaluation of the BDE (3.30) for the assignments $a = 0$ and $b = 0$. Due to the disjunctive form of the left-hand side, this BDE is satisfied for the chosen assignment if one of the conjunctions is equal to 1. Each conjunction can be evaluated separately. Hence, for the first conjunction we have to find a function $f(a,b)$ that satisfies the equation:

$$[f(a,b)]_{a=0,b=0} \wedge \left[\frac{\partial f(a,b)}{\partial a}\right]_{a=0,b=0} \wedge \left[\frac{\partial f(a,b)}{\partial b}\right]_{a=0,b=0} \wedge \left[\frac{\partial f(a,b)}{\partial(a,b)}\right]_{a=0,b=0} = 1 \quad (3.31)$$

which can be transformed into the equivalent system of equations:

$$[f(a, b)]_{a=0, b=0} = 1 \tag{3.32}$$

$$\left[\frac{\partial f(a, b)}{\partial a}\right]_{a=0, b=0} = 1 \tag{3.33}$$

$$\left[\frac{\partial f(a, b)}{\partial b}\right]_{a=0, b=0} = 1 \tag{3.34}$$

$$\left[\frac{\partial f(a, b)}{\partial (a, b)}\right]_{a=0, b=0} = 1 \, . \tag{3.35}$$

The solution of (3.32) is:

$$f(a = 0, b = 0) = 1 \, . \tag{3.36}$$

The second equation (3.33) of the equation system can be solved as follows:

$$\left[\frac{\overline{\partial f(a, b)}}{\partial a}\right]_{a=0, b=0} = 1$$

$$\left[1 \oplus \frac{\partial f(a, b)}{\partial a}\right]_{a=0, b=0} = 1$$

$$[1 \oplus f(a, b) \oplus f(\overline{a}, b)]_{a=0, b=0} = 1$$

$$1 \oplus f(a = 0, b = 0) \oplus f(a = 1, b = 0) = 1$$

$$f(a = 1, b = 0) = f(a = 0, b = 0) \, . \tag{3.37}$$

The known solution (3.36) can be substituted into (3.37) so that the second function value is determined.

$$f(a = 1, b = 0) = 1 \, . \tag{3.38}$$

The third equation (3.34) of the equation system can be solved in the same way using the solution (3.36) in the last step.

$$\left[\frac{\overline{\partial f(a, b)}}{\partial b}\right]_{a=0, b=0} = 1$$

$$\left[1 \oplus \frac{\partial f(a, b)}{\partial b}\right]_{a=0, b=0} = 1$$

$$[1 \oplus f(a, b) \oplus f(a, \overline{b})]_{a=0, b=0} = 1$$

$$1 \oplus f(a = 0, b = 0) \oplus f(a = 0, b = 1) = 1$$

$$f(a = 0, b = 1) = f(a = 0, b = 0)$$

$$f(a = 0, b = 1) = 1 \tag{3.39}$$

The fourth equation (3.35) of the equation system is solved slightly differently due to the missing negation. The substitution of the solution (3.36) in the last step determines the last missing function value of the function $f(a, b)$.

$$\left[\frac{\partial f(a, b)}{\partial (a, b)}\right]_{a=0, b=0} = 1$$
$$[f(a, b) \oplus f(\overline{a}, \overline{b})]_{a=0, b=0} = 1$$
$$f(a = 0, b = 0) \oplus f(a = 1, b = 1) = 1$$
$$f(a = 1, b = 1) = 1 \oplus f(a = 0, b = 0)$$
$$f(a = 1, b = 1) = 1 \oplus 1$$
$$f(a = 1, b = 1) = 0 \tag{3.40}$$

The four function values determined in (3.36), (3.38), (3.39), and (3.40) specify the potential solution function $f(a, b) = \overline{a} \vee \overline{b}$ generated by the first conjunction of the BDE (3.30) and the assignments $a = 0$ and $b = 0$.

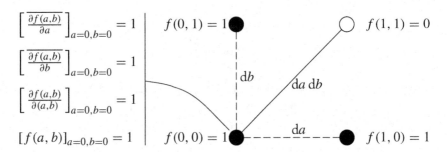

Figure 3.1: Potential solution function $f(a, b) = \overline{a} \vee \overline{b}$ described by (3.31) for $(a, b) = (0, 0)$.

Fig. 3.1 illustrates how the values of $f(a, b)$ and the derivatives with regard to all three directions of the single selected point $a = 0, b = 0$ determine the function values of all other points of the Boolean space. The black filled circles indicate function values 1, and the white filled circle describes a function value 0. The horizontal dashed line means that due to the negated derivative with regard to a the function value $f(0, 0)$ is not changed by changing the value of a. Hence, $f(1, 0)$ is equal to 1. The derivative of $f(a, b)$ with regard to b is negated in the evaluated conjunction, too. This is indicated by the vertical dashed line and determines the function value $f(0, 1) = 1$. The vectorial derivative of $f(a, b)$ with regard to both variables (a, b) appears not negated in the equation (3.31). Hence, there is a change of the function value from $f(0, 0) = 1$ to $f(1, 1) = 0$. Fig. 3.1 depicts this change by a solid line between these two nodes.

In the same way seven further potential solution functions can be built evaluating each of the other seven conjunctions of (3.30). Tab. 3.3 summarizes this complete evaluation of the BDE (3.30) for the assignments $a = 0$ and $b = 0$.

Table 3.3: Potential solution functions $f(a, b)$ for the BDE (3.30) and the assignments $a = 0$ and $b = 0$

conjunction of the BDE					$f(0, 0)$	$f(0, 1)$	$f(1, 0)$	$f(1, 1)$	$f(a, b)$
$f(a, b)$	$\frac{\partial f(a,b)}{\partial a}$	$\overline{\frac{\partial f(a,b)}{\partial b}}$	$\overline{\frac{\partial f(a,b)}{\partial (a,b)}}$		1	1	1	0	$\bar{a} \vee \bar{b}$
$f(a, b)$	$\frac{\partial f(a,b)}{\partial a}$	$\overline{\frac{\partial f(a,b)}{\partial b}}$	$\frac{\partial f(a,b)}{\partial (a,b)}$		1	1	0	1	$\bar{a} \vee b$
$f(a, b)$	$\frac{\partial f(a,b)}{\partial a}$	$\frac{\partial f(a,b)}{\partial b}$	$\overline{\frac{\partial f(a,b)}{\partial (a,b)}}$		1	0	0	1	$a \odot b$
$f(a, b)$	$\overline{\frac{\partial f(a,b)}{\partial a}}$	$\frac{\partial f(a,b)}{\partial b}$	$\frac{\partial f(a,b)}{\partial (a,b)}$		1	0	1	1	$a \vee \bar{b}$
$\overline{f(a, b)}$	$\overline{\frac{\partial f(a,b)}{\partial a}}$	$\overline{\frac{\partial f(a,b)}{\partial b}}$	$\frac{\partial f(a,b)}{\partial (a,b)}$		0	0	0	1	$a \wedge b$
$\overline{f(a, b)}$	$\frac{\partial f(a,b)}{\partial a}$	$\overline{\frac{\partial f(a,b)}{\partial b}}$	$\overline{\frac{\partial f(a,b)}{\partial (a,b)}}$		0	0	1	0	$a \wedge \bar{b}$
$\overline{f(a, b)}$	$\frac{\partial f(a,b)}{\partial a}$	$\frac{\partial f(a,b)}{\partial b}$	$\overline{\frac{\partial f(a,b)}{\partial (a,b)}}$		0	1	1	0	$a \oplus b$
$\overline{f(a, b)}$	$\overline{\frac{\partial f(a,b)}{\partial a}}$	$\frac{\partial f(a,b)}{\partial b}$	$\overline{\frac{\partial f(a,b)}{\partial (a,b)}}$		0	1	0	0	$\bar{a} \wedge b$

    The evaluation of the BDE (3.30) for the selected assignments $a = 0$ and $b = 0$ reveals an important insight into the essence of Boolean differential equations. The function value of $f(a = 0, b = 0)$ is determined by these assignments and the polarity of $f(a, b)$ in a conjunction of the BDE expressed in disjunctive normal form. The polarities of both the function $f(a, b)$ and the simple derivative $\frac{\partial f(a,b)}{\partial a}$ determine the function value of $f(a = 1, b = 0)$. Similarly, the polarities of both the function $f(a, b)$ and the simple derivative $\frac{\partial f(a,b)}{\partial b}$ determine the function value of $f(a = 0, b = 1)$. The remaining function value of $f(a = 1, b = 1)$ is determined by the polarities of both the function $f(a, b)$ and the vectorial derivative $\frac{\partial f(a,b)}{\partial (a,b)}$. Hence, based on a single position in the Boolean space, a conjunction of the function and all simple and vectorial derivatives in certain polarities determine a potential solution function of the BDE (3.30). For that reason the given simple BDE was transformed into a disjunctive normal form such that each conjunction consists of $f(a, b)$ and all simple and vectorial derivatives in a certain polarity.

    No other function as given in the last column of Tab. 3.3 can be a solution of the BDE (3.30) which is equivalent to the BDE (3.25). Different assignments of Boolean values to the variables a and b generate generally different potential solution functions. A sufficient condition for a solution function is that the function is a potential solution function for *each* assignment of Boolean values to the variables a and b. Hence, the evaluation of the BDE (3.30) must be repeated for the assignments $(a, b) = (1, 0)$, $(a, b) = (0, 1)$, and $(a, b) = (1, 1)$. Tabs. 3.4, 3.5, and 3.6 summarize the potential solution functions for the BDE (3.30).

Table 3.4: Potential solution functions $f(a, b)$ for the BDE (3.30) and the assignments $a = 1$ and $b = 0$

conjunction of the BDE				$f(0, 0)$	$f(0, 1)$	$f(1, 0)$	$f(1, 1)$	$f(a, b)$
$f(a, b)$	$\overline{\frac{\partial f(a,b)}{\partial a}}$	$\overline{\frac{\partial f(a,b)}{\partial b}}$	$\frac{\partial f(a,b)}{\partial(a,b)}$	1	0	1	1	$a \vee \overline{b}$
$f(a, b)$	$\frac{\partial f(a,b)}{\partial a}$	$\overline{\frac{\partial f(a,b)}{\partial b}}$	$\frac{\partial f(a,b)}{\partial(a,b)}$	0	1	1	1	$a \vee b$
$f(a, b)$	$\frac{\partial f(a,b)}{\partial a}$	$\frac{\partial f(a,b)}{\partial b}$	$\overline{\frac{\partial f(a,b)}{\partial(a,b)}}$	0	1	1	0	$a \oplus b$
$f(a, b)$	$\overline{\frac{\partial f(a,b)}{\partial a}}$	$\frac{\partial f(a,b)}{\partial b}$	$\overline{\frac{\partial f(a,b)}{\partial(a,b)}}$	1	1	1	0	$\overline{a} \vee \overline{b}$
$\overline{f(a, b)}$	$\frac{\partial f(a,b)}{\partial a}$	$\overline{\frac{\partial f(a,b)}{\partial b}}$	$\frac{\partial f(a,b)}{\partial(a,b)}$	0	1	0	0	$\overline{a} \wedge b$
$\overline{f(a, b)}$	$\frac{\partial f(a,b)}{\partial a}$	$\overline{\frac{\partial f(a,b)}{\partial b}}$	$\frac{\partial f(a,b)}{\partial(a,b)}$	1	0	0	0	$\overline{a} \wedge \overline{b}$
$\overline{f(a, b)}$	$\frac{\partial f(a,b)}{\partial a}$	$\frac{\partial f(a,b)}{\partial b}$	$\overline{\frac{\partial f(a,b)}{\partial(a,b)}}$	1	0	0	1	$a \odot b$
$\overline{f(a, b)}$	$\frac{\partial f(a,b)}{\partial a}$	$\frac{\partial f(a,b)}{\partial b}$	$\frac{\partial f(a,b)}{\partial(a,b)}$	0	0	0	1	$a \wedge b$

The real solution functions of the BDE (3.30) can be identified based on the sets of potential solution functions for each assignment of constant Boolean values to the variables a and b. The intersection of the function sets of Tables 3.3 and 3.4 for the assignments $(a, b) = (0, 0)$ and $(a, b) = (1, 0)$ is

$$\{\overline{a} \vee \overline{b}, a \odot b, a \vee \overline{b}, a \wedge b, a \oplus b, \overline{a} \wedge b\} \ .$$

The intersection of this function set with the set of potential solution functions of the BDE (3.30) for the assignment $(a, b) = (0, 1)$ as shown in Tab. 3.5 reduces the set of intermediate solution functions to

$$\{\overline{a} \vee \overline{b}, a \odot b, a \wedge b, a \oplus b\} \ .$$

The real solution functions of the BDE (3.30) are the functions which belong to the intersection of this intermediate function set and the set of potential solution functions of the BDE (3.30) for the assignment $(a, b) = (1, 1)$ as shown in Tab. 3.6:

$$\{a \odot b, a \oplus b\} \ . \tag{3.41}$$

These two functions are also the solutions of the initial BDE (3.25) due to the equivalent transformations. The substitution of these functions into the BDE (3.25), the calculation of the derivatives followed by a simplification verifies that $f(a, b) = a \odot b = 1 \oplus a \oplus b$ and $f(a, b) =$

Table 3.5: Potential solution functions $f(a, b)$ for the BDE (3.30) and the assignments $a = 0$ and $b = 1$

conjunction of the BDE	$f(0, 0)$	$f(0, 1)$	$f(1, 0)$	$f(1, 1)$	$f(a, b)$
$f(a,b) \quad \dfrac{\partial f(a,b)}{\partial a} \quad \dfrac{\partial f(a,b)}{\partial b} \quad \overline{\dfrac{\partial f(a,b)}{\partial (a,b)}}$	1	1	0	1	$\overline{a} \vee b$
$f(a,b) \quad \dfrac{\partial f(a,b)}{\partial a} \quad \dfrac{\partial f(a,b)}{\partial b} \quad \dfrac{\partial f(a,b)}{\partial (a,b)}$	1	1	1	0	$\overline{a} \vee \overline{b}$
$f(a,b) \quad \dfrac{\partial f(a,b)}{\partial a} \quad \dfrac{\partial f(a,b)}{\partial b} \quad \overline{\dfrac{\partial f(a,b)}{\partial (a,b)}}$	0	1	1	0	$a \oplus b$
$f(a,b) \quad \overline{\dfrac{\partial f(a,b)}{\partial a}} \quad \dfrac{\partial f(a,b)}{\partial b} \quad \dfrac{\partial f(a,b)}{\partial (a,b)}$	0	1	1	1	$a \vee b$
$\overline{f(a,b)} \quad \dfrac{\partial f(a,b)}{\partial a} \quad \dfrac{\partial f(a,b)}{\partial b} \quad \dfrac{\partial f(a,b)}{\partial (a,b)}$	0	0	1	0	$a \wedge \overline{b}$
$\overline{f(a,b)} \quad \dfrac{\partial f(a,b)}{\partial a} \quad \dfrac{\partial f(a,b)}{\partial b} \quad \dfrac{\partial f(a,b)}{\partial (a,b)}$	0	0	0	1	$a \wedge b$
$\overline{f(a,b)} \quad \dfrac{\partial f(a,b)}{\partial a} \quad \dfrac{\partial f(a,b)}{\partial b} \quad \overline{\dfrac{\partial f(a,b)}{\partial (a,b)}}$	1	0	0	1	$a \odot b$
$\overline{f(a,b)} \quad \dfrac{\partial f(a,b)}{\partial a} \quad \dfrac{\partial f(a,b)}{\partial b} \quad \dfrac{\partial f(a,b)}{\partial (a,b)}$	1	0	0	0	$\overline{a} \wedge \overline{b}$

$a \oplus b$ are solution functions of the BDE (3.25). For the first solution function we get:

$$\frac{\partial f(a, b)}{\partial a} \vee \frac{\partial f(a, b)}{\partial b} = \overline{\frac{\partial f(a, b)}{\partial (a, b)}}$$

$$(f(a, b) \oplus f(\overline{a}, b)) \vee (f(a, b) \oplus f(a, \overline{b})) = 1 \oplus f(a, b) \oplus f(\overline{a}, \overline{b})$$

$$(1 \oplus a \oplus b \oplus 1 \oplus \overline{a} \oplus b) \vee (1 \oplus a \oplus b \oplus 1 \oplus a \oplus \overline{b}) \overset{?}{=} 1 \oplus 1 \oplus a \oplus b \oplus 1 \oplus \overline{a} \oplus \overline{b}$$

$$(a \oplus \overline{a}) \vee (b \oplus \overline{b}) \overset{?}{=} a \oplus b \oplus 1 \oplus \overline{a} \oplus \overline{b}$$

$$1 \vee 1 \overset{?}{=} 1 \oplus 1 \oplus 1$$

$$1 = 1 , \tag{3.42}$$

and the following equivalent transformation verifies that the second function $f(a, b) = a \oplus b$ satisfies the BDE (3.25) too:

$$\frac{\partial f(a, b)}{\partial a} \vee \frac{\partial f(a, b)}{\partial b} = \overline{\frac{\partial f(a, b)}{\partial (a, b)}}$$

$$(f(a, b) \oplus f(\overline{a}, b)) \vee (f(a, b) \oplus f(a, \overline{b})) = 1 \oplus f(a, b) \oplus f(\overline{a}, \overline{b})$$

$$(a \oplus b \oplus \overline{a} \oplus b) \vee (a \oplus b \oplus a \oplus \overline{b}) \overset{?}{=} 1 \oplus a \oplus b \oplus \overline{a} \oplus \overline{b}$$

$$(a \oplus \overline{a}) \vee (b \oplus \overline{b}) \overset{?}{=} 1 \oplus a \oplus b \oplus \overline{a} \oplus \overline{b}$$

$$1 \vee 1 \overset{?}{=} 1 \oplus 1 \oplus 1$$

Table 3.6: Potential solution functions $f(a, b)$ for the BDE (3.30) and the assignments $a = 1$ and $b = 1$

conjunction of the BDE				$f(0,0)$	$f(0,1)$	$f(1,0)$	$f(1,1)$	$f(a,b)$
$f(a,b)$	$\frac{\partial f(a,b)}{\partial a}$	$\frac{\partial f(a,b)}{\partial b}$	$\frac{\partial f(a,b)}{\partial (a,b)}$	0	1	1	1	$a \vee b$
$f(a,b)$	$\frac{\partial f(a,b)}{\partial a}$	$\overline{\frac{\partial f(a,b)}{\partial b}}$	$\overline{\frac{\partial f(a,b)}{\partial (a,b)}}$	1	0	1	1	$a \vee \overline{b}$
$f(a,b)$	$\frac{\partial f(a,b)}{\partial a}$	$\frac{\partial f(a,b)}{\partial b}$	$\overline{\frac{\partial f(a,b)}{\partial (a,b)}}$	1	0	0	1	$a \odot b$
$f(a,b)$	$\overline{\frac{\partial f(a,b)}{\partial a}}$	$\frac{\partial f(a,b)}{\partial b}$	$\overline{\frac{\partial f(a,b)}{\partial (a,b)}}$	1	1	0	1	$\overline{a} \vee b$
$\overline{f(a,b)}$	$\overline{\frac{\partial f(a,b)}{\partial a}}$	$\overline{\frac{\partial f(a,b)}{\partial b}}$	$\frac{\partial f(a,b)}{\partial (a,b)}$	1	0	0	0	$\overline{a} \wedge \overline{b}$
$\overline{f(a,b)}$	$\frac{\partial f(a,b)}{\partial a}$	$\overline{\frac{\partial f(a,b)}{\partial b}}$	$\frac{\partial f(a,b)}{\partial (a,b)}$	0	1	0	0	$\overline{a} \wedge b$
$\overline{f(a,b)}$	$\frac{\partial f(a,b)}{\partial a}$	$\frac{\partial f(a,b)}{\partial b}$	$\overline{\frac{\partial f(a,b)}{\partial (a,b)}}$	0	1	1	0	$a \oplus b$
$\overline{f(a,b)}$	$\overline{\frac{\partial f(a,b)}{\partial a}}$	$\frac{\partial f(a,b)}{\partial b}$	$\overline{\frac{\partial f(a,b)}{\partial (a,b)}}$	0	0	1	0	$a \wedge \overline{b}$

$$1 = 1 \, . \tag{3.43}$$

The function $f(a, b) = a \wedge b$ is a potential solution function of the BDE (3.25) for $(a, b) = (0, 0)$, $(a, b) = (1, 0)$, $(a, b) = (0, 1)$, but not for $(a, b) = (1, 1)$. This can be seen as follows:

$$\frac{\partial f(a, b)}{\partial a} \vee \frac{\partial f(a, b)}{\partial b} = \overline{\frac{\partial f(a, b)}{\partial (a, b)}}$$
$$(f(a, b) \oplus f(\overline{a}, b)) \vee (f(a, b) \oplus f(a, \overline{b})) = 1 \oplus f(a, b) \oplus f(\overline{a}, \overline{b})$$
$$(a\,b \oplus \overline{a}\,b) \vee (a\,b \oplus a\,\overline{b}) \stackrel{?}{=} 1 \oplus a\,b \oplus \overline{a}\,\overline{b}$$
$$b \cdot (a \oplus \overline{a}) \vee a \cdot (b \oplus \overline{b}) \stackrel{?}{=} a\,b \oplus (a \vee b)$$
$$b \vee a \stackrel{?}{=} a\,b \oplus a \oplus b \oplus a\,b$$
$$a \vee b \neq a \oplus b \, . \tag{3.44}$$

3.4.2 CLASSES OF SOLUTION FUNCTIONS

From Sec. 3.2 we know that the solution of a BDE is a set of Boolean functions. The set of solution functions contains in the special case of the BDE (3.25) two functions (3.41). Both of them are linear functions which depend on the variables a and b. This observation raises the question about the general structure of solution sets of BDEs of the type (3.22). The answer to this question helps to find an efficient solution algorithm of BDEs (3.22). We answer this question in this subsection.

The following simple example gives a comprehensible insight into the theoretical background of the solution sets of BDEs. Assume, the function

$$f(a, b) = a \vee b \tag{3.45}$$

is a solution function of a BDE of the type (3.22). Fig. 3.2 depicts in the upper part the values of this function in a Boolean space B^2. The function values are annotated on the nodes of this space. The function values 1 are depicted by circles filled with black color. The empty circle indicates the single function value 0.

Next, it can be evaluated for each node of the Boolean space and each direction of change whether the opposite function value is reached. Each solid line in the Boolean space of Fig. 3.2 connects nodes of different function values. Hence, the associated derivatives must be equal to 1 for such nodes of the Boolean space. Vice versa, each dashed line in the Boolean space of Fig. 3.2 connects nodes of identical function values. The derivative must be equal to 0 for such nodes where a dash line begins in the associated direction of change.

Using the known function value and the knowledge about the change of the function value for each direction, simple characteristic equations in which on the left-hand side either the function or a derivative of the function in a certain polarity occurs can be built. There are 2^n such simple equations for each node of the Boolean space B^n. Fig. 3.2 shows in the middle part for each of the four nodes of B^2 the four equations which describe the complete knowledge taken from the associated node.

Each of these four systems of equations can be transformed into an equivalent single equation:

$$f(a, b) \wedge \overline{\frac{\partial f(a, b)}{\partial a}} \wedge \frac{\partial f(a, b)}{\partial b} \wedge \overline{\frac{\partial f(a, b)}{\partial(a, b)}} = 1 , \tag{3.46}$$

$$\overline{f(a, b)} \wedge \frac{\partial f(a, b)}{\partial a} \wedge \frac{\partial f(a, b)}{\partial b} \wedge \frac{\partial f(a, b)}{\partial(a, b)} = 1 , \tag{3.47}$$

$$f(a, b) \wedge \frac{\partial f(a, b)}{\partial a} \wedge \overline{\frac{\partial f(a, b)}{\partial b}} \wedge \frac{\partial f(a, b)}{\partial(a, b)} = 1 , \tag{3.48}$$

$$f(a, b) \wedge \overline{\frac{\partial f(a, b)}{\partial a}} \wedge \overline{\frac{\partial f(a, b)}{\partial b}} \wedge \frac{\partial f(a, b)}{\partial(a, b)} = 1 . \tag{3.49}$$

Fig. 3.2 shows that the function (3.45) satisfies each of the four BDEs (3.46), (3.47), (3.48), and (3.49) only for one assignment of constant values to the variables a and b. Hence, a disjunction of the four BDEs (3.46), (3.47), (3.48), and (3.49) is necessary in a characteristic BDE for the solution function (3.45):

$$
\begin{aligned}
&f(a, b) \wedge \overline{\frac{\partial f(a, b)}{\partial a}} \wedge \frac{\partial f(a, b)}{\partial b} \wedge \overline{\frac{\partial f(a, b)}{\partial(a, b)}} \\
\vee &\overline{f(a, b)} \wedge \frac{\partial f(a, b)}{\partial a} \wedge \frac{\partial f(a, b)}{\partial b} \wedge \frac{\partial f(a, b)}{\partial(a, b)} \\
\vee &f(a, b) \wedge \frac{\partial f(a, b)}{\partial a} \wedge \overline{\frac{\partial f(a, b)}{\partial b}} \wedge \frac{\partial f(a, b)}{\partial(a, b)}
\end{aligned}
$$

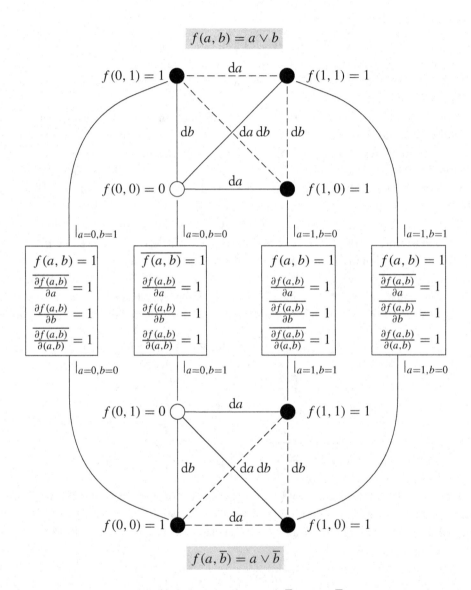

Figure 3.2: Construction of the solution function $f(a, \overline{b}) = a \vee \overline{b}$ from a known solution function $f(a, b) = a \vee b$.

$$\vee f(a,b) \wedge \frac{\overline{\partial f(a,b)}}{\partial a} \wedge \frac{\overline{\partial f(a,b)}}{\partial b} \wedge \frac{\partial f(a,b)}{\partial (a,b)} = 1 .$$
(3.50)

Each of the four BDEs (3.46), (3.47), (3.48), and (3.49) describes for each of the four nodes of B^2 one potential solution function. At the bottom of Fig. 3.2 the systems of differential equations are used for other nodes of the Boolean space. The chosen assignments of the systems of BDEs to the nodes of the Boolean space are labeled on the connection lines in Fig. 3.2. It can be seen that in this example the assignment of the variable a remains unchanged for both the upper and the lower Boolean space, and the assignment of the variable b is exchanged for all for assignments between the upper and the lower Boolean space.

Fig. 3.2 shows in the lower Boolean space that the function $f(a,\overline{b}) = a \vee \overline{b}$ is created in consequence of the exchange of the assignment of constant values to the variable b. This observation does not depend on the explored function (3.45). The exchange is also not restricted to the variable b, but can be executed for each variable. Hence, if $f(a,b)$ is a solution function of a BDE of the type (3.22), then the functions $f(\overline{a},b)$, $f(a,\overline{b})$, and $f(\overline{a},\overline{b})$ are solution functions of the same BDE.

Th. 3.1 generalizes the explored example and states that either all Boolean functions of a certain set or not any of these Boolean functions belong to the solution set of a BDE (3.22).

Theorem 3.1   *If the Boolean function $f(\mathbf{x}) = f(x_1, x_2, ..., x_n)$ is a solution function of the BDE (3.22), then all functions*

$$f(x_1, x_2, ..., x_n) = f(x_1 \oplus c_1, x_2 \oplus c_2, ..., x_n \oplus c_n)$$
(3.51)

for $\mathbf{c} = (c_1, \ldots, c_n) \in B^n$ are solution functions of (3.22) too.

Proof. The term $x_i \oplus c_i$ can be expressed by:

$$x_i \oplus c_i = \begin{cases} x_i & : & c_i = 0 \\ \overline{x}_i & : & c_i = 1 \end{cases}$$
(3.52)

1. Using the Shannon decompositions

$$\begin{aligned} f(\mathbf{x}_0, x_i, \mathbf{x}_1) &= \overline{x}_i \, f(\mathbf{x}_0, x_i = 0, \mathbf{x}_1) \oplus x_i \, f(\mathbf{x}_0, x_i = 1, \mathbf{x}_1) \,, \\ f(\mathbf{x}_0, x_i, \mathbf{x}_1) &= \overline{x}_i \, f(\mathbf{x}_0, 0, \mathbf{x}_1) \oplus x_i \, f(\mathbf{x}_0, 1, \mathbf{x}_1) \,, \\ f(\mathbf{x}_0, \overline{x}_i, \mathbf{x}_1) &= \overline{x}_i \, f(\mathbf{x}_0, \overline{x}_i = 0, \mathbf{x}_1) \oplus x_i \, f(\mathbf{x}_0, \overline{x}_i = 1, \mathbf{x}_1) \,, \\ f(\mathbf{x}_0, \overline{x}_i, \mathbf{x}_1) &= \overline{x}_i \, f(\mathbf{x}_0, x_i = 1, \mathbf{x}_1) \oplus x_i \, f(\mathbf{x}_0, x_i = 0, \mathbf{x}_1) \,, \\ f(\mathbf{x}_0, \overline{x}_i, \mathbf{x}_1) &= \overline{x}_i \, f(\mathbf{x}_0, 1, \mathbf{x}_1) \oplus x_i \, f(\mathbf{x}_0, 0, \mathbf{x}_1) \,, \end{aligned}$$
(3.53)

(3.54)

it follows that

$$f(\mathbf{x}_0, x_i, \mathbf{x}_1)|_{(\mathbf{x}_0, x_i, \mathbf{x}_1) = (\mathbf{c}_0, c_i, \mathbf{c}_1)} = f(\mathbf{x}_0, \overline{x}_i, \mathbf{x}_1)|_{(\mathbf{x}_0, x_i, \mathbf{x}_1) = (\mathbf{c}_0, \overline{c}_i, \mathbf{c}_1)} , \quad (3.55)$$

$$\frac{\partial f(\mathbf{x}_0, x_i, \mathbf{x}_1)}{\partial \mathbf{x}_0}\bigg|_{(\mathbf{x}_0, x_i, \mathbf{x}_1) = (\mathbf{c}_0, c_i, \mathbf{c}_1)} = \frac{\partial f(\mathbf{x}_0, \overline{x}_i, \mathbf{x}_1)}{\partial \mathbf{x}_0}\bigg|_{(\mathbf{x}_0, x_i, \mathbf{x}_1) = (\mathbf{c}_0, \overline{c}_i, \mathbf{c}_1)} , \quad (3.56)$$

$$\frac{\partial f(\mathbf{x}_0, x_i, \mathbf{x}_1)}{\partial (\mathbf{x}_0, x_i)}\bigg|_{(\mathbf{x}_0, x_i, \mathbf{x}_1) = (\mathbf{c}_0, c_i, \mathbf{c}_1)} = \frac{\partial f(\mathbf{x}_0, \overline{x}_i, \mathbf{x}_1)}{\partial (\mathbf{x}_0, x_i)}\bigg|_{(\mathbf{x}_0, x_i, \mathbf{x}_1) = (\mathbf{c}_0, \overline{c}_i, \mathbf{c}_1)} , \quad (3.57)$$

and therefore, (3.51) is true for $\mathbf{c} = (\mathbf{c}_0, c_i, \mathbf{c}_1) = (\mathbf{0}, 1, \mathbf{0})$.

2. All 2^n vectors $\mathbf{c}$ can be created by the antivalence $\oplus$ of vectors with $|\mathbf{c}| = 1$. Th. 3.1 is completely proven when step 1 is applied 2^n times for different solution functions in each iteration.

$$\square$$

All functions that satisfy (3.51) belong to an important relation R_c as defined in (3.58).

Definition 3.2   The set of all 2^{2^n} Boolean functions $f_i(\mathbf{x}) = f_i(x_1, x_2, \ldots, x_n) : B^n \to B$ is called A. The relation $R_c \subseteq A \times A$ is defined by

$$R_c = \left\{ (f_1(\mathbf{x}), f_2(\mathbf{x})) | f_1(\mathbf{x}) \in A, f_2(\mathbf{x}) = f_1(x_1 \oplus c_1, x_2 \oplus c_2, \ldots, x_n \oplus c_n), \forall \mathbf{c} \in B^n \right\} \quad (3.58)$$

Theorem 3.3   *The relation R_c (3.58) is an equivalence relation.*

Proof. An equivalence relation holds the properties reflexivity, symmetry, and transitivity.

A reflexive relation is a binary relation in which each element is related to itself. For $\mathbf{c} = \mathbf{0}$ we get from (3.58)

$$f_2(\mathbf{x}) = f_1(x_1 \oplus 0, x_2 \oplus 0, \ldots, x_n \oplus 0) ,$$
$$f_2(\mathbf{x}) = f_1(x_1, x_2, \ldots, x_n) ,$$
$$f_2(\mathbf{x}) = f_1(\mathbf{x}) , \quad (3.59)$$

which proves the reflexivity.

A symmetric relation is a binary relation in which for each pair of elements $f_1(\mathbf{x}) \in A$ and $f_2(\mathbf{x}) \in A$ holds the condition that if $f_1(\mathbf{x})$ is related to $f_2(\mathbf{x})$ then $f_2(\mathbf{x})$ is related to $f_1(\mathbf{x})$. For an arbitrary constant vector $\mathbf{c} = (c_1, c_2, \ldots, c_n)$ the function $f_2(\mathbf{x})$ is built based on $f_1(\mathbf{x})$ by

$$f_2(x_1, x_2, \ldots, x_n) = f_1(x_1 \oplus c_1, x_2 \oplus c_2, \ldots, x_n \oplus c_n) . \quad (3.60)$$

If $(f_1(\mathbf{x}), f_2(\mathbf{x})) \in R_c$ with $f_2(\mathbf{x})$ defined by (3.60) then $(f_2(\mathbf{x}), f_1(\mathbf{x})) \in R_c$ because the same constant vector $\mathbf{c} = (c_1, c_2, \ldots, c_n)$ changes the first element $f_2(\mathbf{x})$ into

$$f_2(x_1 \oplus c_1, x_2 \oplus c_2, \ldots, x_n \oplus c_n) = f_1(x_1 \oplus c_1 \oplus c_1, x_2 \oplus c_2 \oplus c_2, \ldots, x_n \oplus c_n \oplus c_n) ,$$
$$= f_1(x_1, x_2, \ldots, x_n) , \tag{3.61}$$

and due to (3.60) the second element $f_1(x_1, x_2, \ldots, x_n)$ into $f_2(x_1, x_2, \ldots, x_n)$ which proves the symmetry.

A binary relation over the set A is transitive if the relation holds

$$\forall f_1(\mathbf{x}), f_2(\mathbf{x}), f_3(\mathbf{x}) \in A :$$
$$[(f_1(\mathbf{x}), f_2(\mathbf{x})) \in R_c \wedge (f_2(\mathbf{x}), f_3(\mathbf{x})) \in R_c] \Rightarrow (f_1(\mathbf{x}), f_3(\mathbf{x})) \in R_c . \tag{3.62}$$

Two different constant vectors $^{12}\mathbf{c} = (^{12}c_1, {}^{12}c_2, \ldots, {}^{12}c_n)$ and $^{23}\mathbf{c} = (^{23}c_1, {}^{23}c_2, \ldots, {}^{23}c_n)$ can be used to describe the function of the premise of (3.62):

$$f_2(x_1, x_2, \ldots, x_n) = f_1(x_1 \oplus {}^{12}c_1, x_2 \oplus {}^{12}c_2, \ldots, x_n \oplus {}^{12}c_n) , \tag{3.63}$$
$$f_3(x_1, x_2, \ldots, x_n) = f_2(x_1 \oplus {}^{23}c_1, x_2 \oplus {}^{23}c_2, \ldots, x_n \oplus {}^{23}c_n) . \tag{3.64}$$

The equations (3.63) and (3.64) can be merged into

$$f_3(x_1, x_2, \ldots, x_n) = f_1(x_1 \oplus {}^{12}c_1 \oplus {}^{23}c_1, x_2 \oplus {}^{12}c_2 \oplus {}^{23}c_2, \ldots, x_n \oplus {}^{12}c_n \oplus {}^{23}c_n) ,$$
$$= f_1(x_1 \oplus {}^{13}c_1, x_2 \oplus {}^{13}c_2, \ldots, x_n \oplus {}^{13}c_n) , \tag{3.65}$$

so that the conclusion of (3.62) is satisfied for a constant vector $^{13}\mathbf{c} = {}^{12}\mathbf{c} \oplus {}^{23}\mathbf{c}$ which proves the transitivity. Hence, Theorem 3.3 is completely proven.   □

From Theorem 3.3 we get some useful corollaries.

1. The equivalence relation (3.58) separates the set of all 2^{2^n} Boolean functions $f(\mathbf{x})$ into disjoint subclasses.

2. All functions which relate to each other based on the equation (3.58) constitute a complete equivalence class.

3. The solution set of each Boolean differential equation (3.22) consists of $k \geq 0$ equivalence classes of Boolean functions as specified by (3.58).

Tab. 3.7 shows in detail the allocation of all 16 functions $f(a, b)$ of B^2 to seven equivalence classes. It can be seen that at most $2^2 = 4$ functions belong to one class and each class includes a power of 2 different functions.

The number of function classes in the Boolean space B^3 is equal to 46. Tab. 3.8 shows how many classes contain the same power of 2 different functions. The maximal number of functions in a single equivalence class in the Boolean space B^3 is equal to the number of different Boolean vectors of the length 3 that is $2^3 = 8$ for B^3.

Table 3.7: Complete enumeration of classes of all 16 Boolean functions $f(a, b)$ of B^2

class number	number of functions	functions
1	1	$f(a, b) = 0$
2	1	$f(a, b) = 1$
3	2	$f(a, b) = a$, $f(a, b) = \overline{a}$
4	2	$f(a, b) = b$, $f(a, b) = \overline{b}$
5	2	$f(a, b) = a \oplus b$, $f(a, b) = a \odot b$
6	4	$f(a, b) = a \wedge b$, $f(a, b) = \overline{a} \wedge b$ $f(a, b) = a \wedge \overline{b}$, $f(a, b) = \overline{a} \wedge \overline{b}$
7	4	$f(a, b) = a \vee b$, $f(a, b) = \overline{a} \vee b$ $f(a, b) = a \vee \overline{b}$, $f(a, b) = \overline{a} \vee \overline{b}$

Table 3.8: All 46 classes of all 256 Boolean functions of B^3

number of classes	number of functions within a single class	sum of functions
2	1	2
7	2	14
14	4	56
23	8	184

3.4.3   SEPARATION OF FUNCTION CLASSES

A very efficient solution algorithm for BDEs (3.22) was developed in the Ph.D. thesis of Steinbach [1981] based on the theoretical background explored in Subsecs. 3.4.1 and 3.4.2. The key idea of this algorithm is the separation of the set of solution functions of a BDE (3.22) out of the set of all local solutions of this BDE. We explain this algorithm in a comprehensive manner, and define the required terms for a compact presentation.

Definition 3.4   Let $f(\mathbf{x})$ be a solution function of a BDE (3.22). Then

$$LS = \left[f(\mathbf{x}), \frac{\partial f(\mathbf{x})}{\partial x_1}, \frac{\partial f(\mathbf{x})}{\partial x_2}, ..., \frac{\partial f(\mathbf{x})}{\partial \mathbf{x}} \right]_{\mathbf{x}=\mathbf{c}} \tag{3.66}$$

is a *local solution* for $\mathbf{x} = \mathbf{c}$, and

$$\nabla f(\mathbf{x}) = \left(f(\mathbf{x}), \frac{\partial f(\mathbf{x})}{\partial x_1}, \frac{\partial f(\mathbf{x})}{\partial x_2}, ..., \frac{\partial f(\mathbf{x})}{\partial \mathbf{x}} \right) \tag{3.67}$$

is a short notation for the vector of $f(\mathbf{x})$ and all simple and vectorial derivatives of this Boolean function.

A local solution (3.66) restricts the knowledge to one point $\mathbf{x} = \mathbf{c}$ of the 2^n points of B^n. The local solution is a vector of 2^n Boolean values that determine the function value $f(\mathbf{c})$ and the values of all simple and vectorial derivatives of the same function at the same point $\mathbf{x} = \mathbf{c}$. The information of a single local solution is sufficient to reconstruct the complete function $f(\mathbf{x})$, because the local function value and its changes in all directions are known.

There are two ways to find local solutions of a BDE (3.22).

1. Each Boolean function $f(\mathbf{x})$ can be substituted into the BDE (3.22). Using the definitions of the derivatives and the rules of the Boolean Algebra, it can be verified whether the function $f(\mathbf{x})$ is a solution function. Local solutions can be created based on the definition (3.66) for a known solution function $f(\mathbf{x})$ and each of the 2^n points of B^n. Local solutions found in this way describe at least one solution function of the BDE (3.22).

2. The BDE (3.22) contains only the elements of $\nabla f(\mathbf{x})$ defined by (3.67) either in non-negated or negated form. Hence, we can model these elements of the BDE (3.22) by Boolean variables. The solution of a Boolean equation associated to the BDE (3.22) consists of Boolean vectors with the same structure as local solutions (3.66). The set of local solutions found in this way describes all solution functions of the BDE (3.22) which must be solved, but can contain local solutions of functions which do not belong to the solution set of the BDE (3.22).

The utilization of the first approach requires the successful verification that a chosen function $f(\mathbf{x})$ satisfies the BDE (3.22). A drawback of this approach is that all 2^{2^n} Boolean functions of B^n must be verified in the case that the BDE (3.22) has an empty solution set.

The benefit of the second way is that such a check is not necessary. Local solutions of all potential solution functions of the BDE (3.22) can be calculated directly as solution of an associated Boolean equation of the BDE. This associated Boolean equation must be defined precisely.

Definition 3.5 Based on the Boolean differential equation (3.22) we define the associated Boolean equation

$$D_1(u_0, u_1, ..., u_{2^n-1}) = D_2(u_0, u_1, ..., u_{2^n-1}) . \qquad (3.68)$$

The index i of the variable u_i indicates the associated element of the BDE (3.22) such that a value 1 in the binary code of the index determines a variable in the change vector of the associated derivative. Using this rule, we map

- the function $\quad f(\mathbf{x})$ to u_0 ,

- the derivative $\quad \dfrac{\partial f(\mathbf{x})}{\partial x_1}$ to u_1 ,

- the derivative $\quad \dfrac{\partial f(\mathbf{x})}{\partial x_2}$ to u_2 ,

- the derivative $\dfrac{\partial f(\mathbf{x})}{\partial(x_1,x_2)}$ to u_3 ,

- ..., and finally

- the derivative $\dfrac{\partial f(\mathbf{x})}{\partial \mathbf{x}}$ to u_{2^n-1} .

The solution of the associated Boolean equation (3.68) is the *set of local solutions* SLS.

Each vector $\mathbf{u}$ of the set SLS describes a combination of Boolean values of the function $f(\mathbf{x})$ and all their simple and vectorial derivatives that satisfy the BDE (3.22) at least for one point $\mathbf{x} = \mathbf{c}$ of the Boolean space B^n. A zero value in such a local solution indicates that the associated element must be negated.

Knowing the set of local solutions SLS of the associated Boolean equation (3.68), it remains the task to separate local solutions of actual solution functions of the BDE (3.22) from the other local solutions. It is a necessary condition for each potential solution function $f(\mathbf{x})$ of the BDE (3.22) that at least one local solution belongs to the set of local solutions SLS. The Boolean function $f(\mathbf{x})$ is a solution function of the BDE (3.22) if the necessary and sufficient condition (3.69) becomes true:

$$\forall \mathbf{c} \in B^n \quad \nabla f(\mathbf{x}) \, |_{\mathbf{x}=\mathbf{c}} \in SLS \ . \tag{3.69}$$

The correctness of condition (3.69) follows from the property that a solution function has local solutions for each point of B^n.

From (3.69) we can derive a rule how to extract actual solution functions of the BDE (3.22) out of the set of potential solution functions SLS. If the expression

$$\exists \mathbf{c}' \in B^n \quad \nabla f(\mathbf{x}) \, |_{\mathbf{x}=\mathbf{c}'} \notin SLS \tag{3.70}$$

becomes true, the function $f(\mathbf{x})$ described by the local solution $\nabla f(\mathbf{x}) \, |_{\mathbf{x}=\mathbf{c}'}$ does not solve (3.68), and consequently, this function does not belong to the solution set of the BDE (3.22). The conclusion from both (3.69) and (3.70) is that 2^n local solutions are necessary so that one function $f(\mathbf{x})$ is an element of the solution set of the BDE (3.22). Not all of these local solutions must be different.

It is possible to utilize conditions (3.69) and (3.70) directly to separate the solution function of a BDE (3.22) from all other functions. However, the different meaning of the Boolean values in a local solution complicates such a procedure. It is easier to manipulate function values instead of a single function value and values of the derivatives. For that reason, the values of all derivatives are translated in a simple supplementary step into function values of the reached points of the Boolean space.

The vector $\mathbf{v}$ contains all values of a Boolean function. Hence, both a vector $\mathbf{v}$ and a vector $\mathbf{u}$ represent the complete information about a Boolean function, only the semantics of their interpretation are different. The system of Boolean equations (3.71) and (3.72) describes the transition between these two types of representation of Boolean functions:

$$v_0 = u_0, \tag{3.71}$$
$$v_i = u_0 \oplus u_i, \quad \text{with} \quad i = 1, 2, ..., 2^n - 1. \tag{3.72}$$

The transformation of the set $SLS(\mathbf{u})$ into the set $S(\mathbf{v})$ is used in the algorithm *separation of function classes*. This supplementary task realizes the function d2v() (*derivative to value*). After the transformation $S(\mathbf{v}) \leftarrow \text{d2v}(SLS(\mathbf{u}))$ the set $S(\mathbf{v})$ still contains local solutions. In further steps the set $S(\mathbf{v})$ can be restricted to actual solution functions of the BDE (3.22) by means of the condition (3.70). For that reason the name S (*solution*) is already used.

The separation procedure within the set $S(\mathbf{v})$ utilizes the property of equation (3.51) that the exchange of x_i and $\overline{x}_i$ does not change the set of solution functions. This change can be implemented by exchanging pairs of function values v_j. The index of the variable x_i controls in (3.73) which pairs of function values v_j must exchange in the set $S(\mathbf{v})$:

$$
\begin{aligned}
v_{(m+2k \cdot 2^{i-1})} &\Longleftrightarrow v_{(m+(2k+1) \cdot 2^{i-1})}, \quad \text{with} \\
i &= 1, 2, ..., n, \\
m &= 0, 1, ..., 2^{i-1} - 1, \\
k &= 0, 1, ..., 2^{n-i} - 1.
\end{aligned}
\tag{3.73}
$$

Tab. 3.9 lists in 32 lines the index pairs defined by (3.73) to solve a BDE (3.22) of up to six variables. The value of i indicates which variable x_i of the desired solution function must change.

A second supplementary function epv() (*exchange pairs of values*) exchanges function values of $S(\mathbf{v})$ using formula (3.73) with regard to a given index i and returns the set $ST(\mathbf{v})$. The set $ST(\mathbf{v})$ can be created for each variable x_i by $ST(\mathbf{v}) \leftarrow \text{epv}(S(\mathbf{v}), i)$.

Using the sets $S(\mathbf{v})$ and $ST(\mathbf{v})$ it is very easy to remove such local solutions from the set $S(\mathbf{v})$ which satisfy (3.70) and do therefore not belong to the solution set of the BDE (3.22). Only such vectors $\mathbf{v}$ which belong to the result of the intersection $S(\mathbf{v}) \leftarrow S(\mathbf{v}) \cap ST(\mathbf{v})$ can satisfy the sufficient condition (3.69) for solution functions of the BDE (3.22). This condition requires that the generation of the set $ST(\mathbf{v})$ by the function $\text{epv}(S(\mathbf{v}), i)$ and the consecutive calculation of the intersection with the set $S(\mathbf{v})$ is executed for each variable x_i iteratively.

Algorithm 1 shows all the steps of the explained solution procedure for BDEs of type (3.22). The BDE to be solved is mapped to an associated Boolean equation and solved with regard to the set of local solutions $SLS(\mathbf{u})$ in line 1. This set is transformed into the set $S(\mathbf{v})$ of function vectors of potential solution functions. The *separation of function classes* which solve the BDE (3.22) is realized in the loop of lines 3 to 6. It is a benefit of this algorithm that all 2^n local solutions of all function classes are evaluated by only n intersections of sets $S(\mathbf{v})$ and $ST(\mathbf{v})$ where $ST(\mathbf{v})$ is built by simple exchanges of all columns in a well-defined manner.

Table 3.9: Index pairs of columns for the exchange of function values

$i = 1$	$i = 2$	$i = 3$	$i = 4$	$i = 5$	$i = 6$
0 ⇔ 1	0 ⇔ 2	0 ⇔ 4	0 ⇔ 8	0 ⇔ 16	0 ⇔ 32
2 ⇔ 3	1 ⇔ 3	1 ⇔ 5	1 ⇔ 9	1 ⇔ 17	1 ⇔ 33
4 ⇔ 5	4 ⇔ 6	2 ⇔ 6	2 ⇔ 10	2 ⇔ 18	2 ⇔ 34
6 ⇔ 7	5 ⇔ 7	3 ⇔ 7	3 ⇔ 11	3 ⇔ 19	3 ⇔ 35
8 ⇔ 9	8 ⇔ 10	8 ⇔ 12	4 ⇔ 12	4 ⇔ 20	4 ⇔ 36
10 ⇔ 11	9 ⇔ 11	9 ⇔ 13	5 ⇔ 13	5 ⇔ 21	5 ⇔ 37
12 ⇔ 13	12 ⇔ 14	10 ⇔ 14	6 ⇔ 14	6 ⇔ 22	6 ⇔ 38
14 ⇔ 15	13 ⇔ 15	11 ⇔ 15	7 ⇔ 15	7 ⇔ 23	7 ⇔ 39
16 ⇔ 17	16 ⇔ 18	16 ⇔ 20	16 ⇔ 24	8 ⇔ 24	8 ⇔ 40
18 ⇔ 19	17 ⇔ 19	17 ⇔ 21	17 ⇔ 25	9 ⇔ 25	9 ⇔ 41
20 ⇔ 21	20 ⇔ 22	18 ⇔ 22	18 ⇔ 26	10 ⇔ 26	10 ⇔ 42
22 ⇔ 23	21 ⇔ 23	19 ⇔ 23	19 ⇔ 27	11 ⇔ 27	11 ⇔ 43
24 ⇔ 25	24 ⇔ 26	24 ⇔ 28	20 ⇔ 28	12 ⇔ 28	12 ⇔ 44
26 ⇔ 27	25 ⇔ 27	25 ⇔ 29	21 ⇔ 29	13 ⇔ 29	13 ⇔ 45
28 ⇔ 29	28 ⇔ 30	26 ⇔ 30	22 ⇔ 30	14 ⇔ 30	14 ⇔ 46
30 ⇔ 31	29 ⇔ 31	27 ⇔ 31	23 ⇔ 31	15 ⇔ 31	15 ⇔ 47
32 ⇔ 33	32 ⇔ 34	32 ⇔ 36	32 ⇔ 40	32 ⇔ 48	16 ⇔ 48
34 ⇔ 35	33 ⇔ 35	33 ⇔ 37	33 ⇔ 41	33 ⇔ 49	17 ⇔ 49
36 ⇔ 37	36 ⇔ 38	34 ⇔ 38	34 ⇔ 42	34 ⇔ 50	18 ⇔ 50
38 ⇔ 39	37 ⇔ 39	35 ⇔ 39	35 ⇔ 43	35 ⇔ 51	19 ⇔ 51
40 ⇔ 41	40 ⇔ 42	40 ⇔ 44	36 ⇔ 44	36 ⇔ 52	20 ⇔ 52
42 ⇔ 43	41 ⇔ 43	41 ⇔ 45	37 ⇔ 45	37 ⇔ 53	21 ⇔ 53
44 ⇔ 45	44 ⇔ 46	42 ⇔ 46	38 ⇔ 46	38 ⇔ 54	22 ⇔ 54
46 ⇔ 47	45 ⇔ 47	43 ⇔ 47	39 ⇔ 47	39 ⇔ 55	23 ⇔ 55
48 ⇔ 49	48 ⇔ 50	48 ⇔ 52	48 ⇔ 56	40 ⇔ 56	24 ⇔ 56
50 ⇔ 51	49 ⇔ 51	49 ⇔ 53	49 ⇔ 57	41 ⇔ 57	25 ⇔ 57
52 ⇔ 53	52 ⇔ 54	50 ⇔ 54	50 ⇔ 58	42 ⇔ 58	26 ⇔ 58
54 ⇔ 55	53 ⇔ 55	51 ⇔ 55	51 ⇔ 59	43 ⇔ 59	27 ⇔ 59
56 ⇔ 57	56 ⇔ 58	56 ⇔ 60	52 ⇔ 60	44 ⇔ 60	28 ⇔ 60
58 ⇔ 59	57 ⇔ 59	57 ⇔ 61	53 ⇔ 61	45 ⇔ 61	29 ⇔ 61
60 ⇔ 61	60 ⇔ 62	58 ⇔ 62	54 ⇔ 62	46 ⇔ 62	30 ⇔ 62
62 ⇔ 63	61 ⇔ 63	59 ⇔ 63	55 ⇔ 63	47 ⇔ 63	31 ⇔ 63

Algorithm 1 Separation of function classes

Require: BDE (3.22) in which the function $f(\mathbf{x})$ depends on n variables
Ensure: set S of Boolean vectors $\mathbf{v} = (v_0, v_1, \ldots, v_{2^n-1})$ that describe substituted in (3.74) the set
  of all solution functions of the BDE (3.22)
1: $SLS(\mathbf{u}) \leftarrow$ solution of the Boolean equation (3.68) associated with BDE (3.22)
2: $S(\mathbf{v}) \leftarrow \text{d2v}(SLS(\mathbf{u}))$
3: **for** $i \leftarrow 1$ to n **do**
4:   $ST(\mathbf{v}) \leftarrow \text{epv}(S(\mathbf{v}), i)$
5:   $S(\mathbf{v}) \leftarrow S(\mathbf{v}) \cap ST(\mathbf{v})$
6: **end for**

The result of Algorithm 1 is the set $S(\mathbf{v})$ which describes the set of all solutions functions of the BDE (3.22). A single vector $\mathbf{v}$ specifies one solution function as follows:

$$
\begin{aligned}
f(x_1, x_2, \ldots, x_n) = \ & \overline{x}_1 \, \overline{x}_2 \ldots \overline{x}_n \, v_0 \oplus \\
& x_1 \, \overline{x}_2 \ldots \overline{x}_n \, v_1 \oplus \\
& \overline{x}_1 \, x_2 \ldots \overline{x}_n \, v_2 \oplus \\
& \ldots \oplus \\
& x_1 \, x_2 \ldots x_n \, v_{2^n-1} \; .
\end{aligned}
\tag{3.74}
$$

The benefit of Algorithm 1 becomes clearly visible in direct comparison to the expensive solution procedure of a BDE (3.22) introduced in Subsec. 3.4.1. For that reason we solve the same BDE (3.25) which depends on functions $f(a, b)$. Algorithm 1 is explained for the variables x_i. We associate the variables of the function $f(a, b)$ in natural order with the variables x_i, $i = 1, 2$: $a \leftrightarrow x_1$ and $b \leftrightarrow x_2$.

Using Def. (3.68) the BDE

$$
\frac{\partial f(a, b)}{\partial a} \vee \frac{\partial f(a, b)}{\partial b} = \overline{\frac{\partial f(a, b)}{\partial (a, b)}}
$$

is mapped onto the associated Boolean equation:

$$
u_1 \vee u_2 = \overline{u}_3 \; .
\tag{3.75}
$$

The solution (3.76) of the Boolean equation (3.75) consists of four Boolean vectors which can be expressed by three ternary vectors.

$$
SLS(\mathbf{u}) =
\begin{array}{ccc}
u_1 & u_2 & u_3 \\
\hline
1 & 0 & 0 \\
- & 1 & 0 \\
0 & 0 & 1
\end{array}
\tag{3.76}
$$

The set $SLS(\mathbf{u})$ (3.76) is the result of line 1 of Algorithm 1.

The transformation of the set of local solutions $SLS(\mathbf{u})$ onto the set of function values $S(\mathbf{v})$ is specified by the system of Boolean equations (3.71) and (3.72). The substitution of (3.71) into (3.72) results in

$$v_i \;=\; v_0 \oplus u_i, \quad \text{with} \quad i = 1, 2, ..., 2^n - 1 . \tag{3.77}$$

Hence for $i \neq 0$, the values v_i are equal to the values of u_i if $v_0 = 0$, and in the opposite case of $v_0 = 1$ we have $v_i = \bar{u}_i$.

$$
S(\mathbf{v}) =
\begin{array}{cccc}
v_0 & v_1 & v_2 & v_3 \\
\hline
0 & 1 & 0 & 0 \\
0 & - & 1 & 0 \\
0 & 0 & 0 & 1 \\
1 & 0 & 1 & 1 \\
1 & - & 0 & 1 \\
1 & 1 & 1 & 0 \\
\hline
\end{array}
\tag{3.78}
$$

The set $S(\mathbf{v})$ (3.78) is the result of line 2 of Algorithm 1.

The separation of the function classes requires for the functions $f(a, b)$ of two variables two iterative exchanges of columns of $S(\mathbf{v})$ into $ST(\mathbf{v})$ followed by intersections of these sets. The pairs of columns which must be exchanged are defined by (3.73). The indices of the columns to be exchanged are listed in Tab. 3.9. For the first variable a of $f(a, b)$ the column v_0 must be exchanged with the column v_1 and the column v_2 must be exchanged with the column v_3. This information is given in the first two rows of the column labeled by $i = 1$ of Tab. 3.9.

$$
ST(\mathbf{v}) =
\begin{array}{cccc}
v_0 & v_1 & v_2 & v_3 \\
\hline
1 & 0 & 0 & 0 \\
- & 0 & 0 & 1 \\
0 & 0 & 1 & 0 \\
0 & 1 & 1 & 1 \\
- & 1 & 1 & 0 \\
1 & 1 & 0 & 1 \\
\hline
\end{array}
\tag{3.79}
$$

The set $ST(\mathbf{v})$ (3.79) is the result of first execution of line 4 of Algorithm 1. The new set $S(\mathbf{v}) \leftarrow S(\mathbf{v}) \cap ST(\mathbf{v})$ (3.80) is built by an intersection as specified in line 5 of Algorithm 1.

$$
S(\mathbf{v}) =
\begin{array}{cccc}
v_0 & v_1 & v_2 & v_3 \\
\hline
0 & 1 & 0 & 0 \\
0 & - & 1 & 0 \\
0 & 0 & 0 & 1 \\
1 & 0 & 1 & 1 \\
1 & - & 0 & 1 \\
1 & 1 & 1 & 0 \\
\hline
\end{array}
\cap
\begin{array}{cccc}
v_0 & v_1 & v_2 & v_3 \\
\hline
1 & 0 & 0 & 0 \\
- & 0 & 0 & 1 \\
0 & 0 & 1 & 0 \\
0 & 1 & 1 & 1 \\
- & 1 & 1 & 0 \\
1 & 1 & 0 & 1 \\
\hline
\end{array}
=
\begin{array}{cccc}
v_0 & v_1 & v_2 & v_3 \\
\hline
0 & 0 & 1 & 0 \\
0 & 1 & 1 & 0 \\
0 & 0 & 0 & 1 \\
1 & 0 & 0 & 1 \\
1 & 1 & 0 & 1 \\
1 & 1 & 1 & 0 \\
\hline
\end{array}
\tag{3.80}
$$

This intersection reduced the number of potential solution functions from eight in (3.78) to six in (3.80).

In the second sweep of the iteration of lines 3 to 6 of Algorithm 1 the separation with regard to the second variable b must be executed. The columns to exchange can be taken from Tab. 3.9; the first two rows of the column $i = 2$ specify the exchanges $v_0 \Leftrightarrow v_2$ and $v_1 \Leftrightarrow v_3$.

$$
ST(\mathbf{v}) =
\begin{array}{cccc}
v_0 & v_1 & v_2 & v_3 \\
\hline
1 & 0 & 0 & 0 \\
1 & 0 & 0 & 1 \\
0 & 1 & 0 & 0 \\
0 & 1 & 1 & 0 \\
0 & 1 & 1 & 1 \\
1 & 0 & 1 & 1 \\
\hline
\end{array}
\tag{3.81}
$$

The set $ST(\mathbf{v})$ (3.81) is the result of second execution of line 4 of Algorithm 1. The final set $S(\mathbf{v}) \leftarrow S(\mathbf{v}) \cap ST(\mathbf{v})$ (3.82) is again built by an intersection as specified in line 5 of Algorithm 1.

$$
S(\mathbf{v}) =
\begin{array}{cccc}
v_0 & v_1 & v_2 & v_3 \\
\hline
0 & 0 & 1 & 0 \\
0 & 1 & 1 & 0 \\
0 & 0 & 0 & 1 \\
1 & 0 & 0 & 1 \\
1 & 1 & 0 & 1 \\
1 & 1 & 1 & 0 \\
\hline
\end{array}
\cap
\begin{array}{cccc}
v_0 & v_1 & v_2 & v_3 \\
\hline
1 & 0 & 0 & 0 \\
1 & 0 & 0 & 1 \\
0 & 1 & 0 & 0 \\
0 & 1 & 1 & 0 \\
0 & 1 & 1 & 1 \\
1 & 0 & 1 & 1 \\
\hline
\end{array}
=
\begin{array}{cccc}
v_0 & v_1 & v_2 & v_3 \\
\hline
0 & 1 & 1 & 0 \\
1 & 0 & 0 & 1 \\
\hline
\end{array}
\tag{3.82}
$$

The set $S(\mathbf{v})$ (3.82) is the result of second execution of line 5 of Algorithm 1. This set describes two solution functions of the BDE (3.25) based on the normal form

$$
f(a, b) = \overline{a}\,\overline{b}\,v_0 \oplus a\,\overline{b}\,v_1 \oplus \overline{a}\,b\,v_2 \oplus a\,b\,v_3
\tag{3.83}
$$

which specializes the equation (3.74). The BDE (3.25) has two solution functions:

$$
\begin{aligned}
f_1(a, b) &= a\,\overline{b} \oplus \overline{a}\,b = a \oplus b \,, \\
f_2(a, b) &= \overline{a}\,\overline{b} \oplus a\,b = a \odot b \,.
\end{aligned}
\tag{3.84}
$$

This is the same set of solution functions as found in Subsection 3.4.1. However, it is much simpler to calculate this solution set of the BDE (3.25) using Algorithm 1.

3.4.4   SEPARATION OF FUNCTION CLASSES USING XBOOLE

The example of Subsection 3.4.3 is so small that it could be solved easily by hand based on Algorithm 1. Larger BDEs require the support of a computer for the calculation of the correct set of solution functions. XBOOLE is a very helpful basic software for all necessary calculation steps of Algorithm 1. There are different possibilities to express Algorithm 1 by means of a problem program (PRP) which can be executed by the XBOOLE-Monitor.

As a first example we show a direct mapping of Algorithm 1 to an XBOOLE-PRP for the BDE (3.25) that was solved in Subsections 3.4.1 and 3.4.3 in order to facilitate a direct comparison. Fig. 3.3 shows this PRP.

1	space 32 1	14	maxk 3 4 5
2	avar 1	15	vtin 1 6
3	u0 u1 u2 u3 v0 v1 v2 v3 .	16	v0 v2 .
4	sbe 1 1	17	vtin 1 7
5	u1+u2 =/ u3 .	18	v1 v3 .
6	sbe 1 2	19	cco 5 6 7 8
7	v0=u0 ,	20	isc 5 8 9
8	v1=u0#u1 ,	21	vtin 1 10
9	v2=u0#u2 ,	22	v0 v1 .
10	v3=u0#u3 .	23	vtin 1 11
11	isc 1 2 3	24	v2 v3 .
12	vtin 1 4	25	cco 9 10 11 12
13	u0 u1 u2 u3 .	26	isc 9 12 13

Figure 3.3: Listing of the PRP to solve the BDE (3.25).

The number in front of the PRP lines has been added in the listing for reference purposes and does not belong to the PRP itself. In line 1 a Boolean space of 32 variables with the number 1 is defined. The used variables are attached to this Boolean space in the wanted order in lines 2 and 3. The associated Boolean equation of the BDE (3.25) is solved in lines 4 and 5 of the listing in Fig. 3.3. These two lines of the PRP realize the task of line 1 in Algorithm 1. The result of this step is the set of local solutions which is stored as object 1 $XBO[1]$ (3.85). The content of this object is equal to $SLS(\mathbf{u})$ shown in (3.76).

$$XBO[1] = SLS(\mathbf{u}) = \begin{array}{ccc} u_1 & u_2 & u_3 \\ \hline 1 & 0 & 0 \\ - & 1 & 0 \\ 0 & 0 & 1 \end{array} \qquad (3.85)$$

The mapping $S(\mathbf{v}) \leftarrow \text{d2v}(SLS(\mathbf{u}))$ of line 2 in Algorithm 1 is realized by lines 6 to 14 of the PRP in Fig. 3.3. The command in line 6 of the PRP solves the set of Boolean equations given in lines 7 to 10 which comply with the system of Boolean equations (3.71) and (3.72) for $n = 2$. The solution set of this system of Boolean equations is the XBOOLE object $XBO[2]$ (3.86) that contains the mapping of all 16 Boolean vectors (u_0, u_1, u_2, u_3) onto the associated vectors (v_0, v_1, v_2, v_3).

$$XBO[2] = \begin{array}{cccccccc} u_0 & u_1 & u_2 & u_3 & v_0 & v_1 & v_2 & v_3 \\ \hline 0 & 1 & 1 & 1 & 0 & 1 & 1 & 1 \\ 0 & 0 & 1 & 1 & 0 & 0 & 1 & 1 \\ 0 & 1 & 0 & 1 & 0 & 1 & 0 & 1 \\ 0 & 0 & 0 & 1 & 0 & 0 & 0 & 1 \\ 1 & 0 & 0 & 0 & 1 & 1 & 1 & 1 \\ 1 & 1 & 0 & 0 & 1 & 0 & 1 & 1 \\ 1 & 0 & 1 & 0 & 1 & 1 & 0 & 1 \\ 1 & 1 & 1 & 0 & 1 & 0 & 0 & 1 \\ 1 & 0 & 0 & 1 & 1 & 1 & 1 & 0 \\ 1 & 1 & 0 & 1 & 1 & 0 & 1 & 0 \\ 1 & 0 & 1 & 1 & 1 & 1 & 0 & 0 \\ 1 & 1 & 1 & 1 & 1 & 0 & 0 & 0 \\ 0 & 1 & 1 & 0 & 0 & 1 & 1 & 0 \\ 0 & 0 & 1 & 0 & 0 & 0 & 1 & 0 \\ 0 & 1 & 0 & 0 & 0 & 1 & 0 & 0 \\ 0 & 0 & 0 & 0 & 0 & 0 & 0 & 0 \\ \hline \end{array} \tag{3.86}$$

The result $XBO[3]$ (3.87) of the intersection in line 11 of this set selects eight vectors $(\mathbf{u}, \mathbf{v})$ which satisfy the local solutions $SLS(\mathbf{u})$ of (3.85).

$$XBO[3] = \begin{array}{cccccccc} u_0 & u_1 & u_2 & u_3 & v_0 & v_1 & v_2 & v_3 \\ \hline 1 & 1 & 0 & 0 & 1 & 0 & 1 & 1 \\ 0 & 1 & 0 & 0 & 0 & 1 & 0 & 0 \\ 1 & 0 & 1 & 0 & 1 & 1 & 0 & 1 \\ 1 & 1 & 1 & 0 & 1 & 0 & 0 & 1 \\ 0 & 1 & 1 & 0 & 0 & 1 & 1 & 0 \\ 0 & 0 & 1 & 0 & 0 & 0 & 1 & 0 \\ 0 & 0 & 0 & 1 & 0 & 0 & 0 & 1 \\ 1 & 0 & 0 & 1 & 1 & 1 & 1 & 0 \\ \hline \end{array} \tag{3.87}$$

The k-fold maximum in line 14 eliminates all columns (u_0, u_1, u_2, u_3) in $XBO[3]$ (3.87). The required variables are prepared as tuple of variables in lines 12 and 13 of the PRP listing in Fig. 3.3. The XBOOLE object $XBO[5]$ (3.88) contains the same eight potential solution vectors as $S(\mathbf{v})$ (3.78). Both the order of the binary vectors and their merge into ternary vectors does not change the meaning of this set.

$$XBO[5] = S(\mathbf{v}) = \begin{array}{cccc} v_0 & v_1 & v_2 & v_3 \\ \hline 1 & 0 & 1 & 1 \\ 0 & 1 & 0 & 0 \\ 1 & 1 & 0 & 1 \\ 1 & 0 & 0 & 1 \\ 0 & 1 & 1 & 0 \\ 0 & 0 & 1 & 0 \\ 0 & 0 & 0 & 1 \\ 1 & 1 & 1 & 0 \\ \hline \end{array} \tag{3.88}$$

It remains the separation of the actual solution functions from the set of potential solution functions (3.88). The solution functions of the BDE (3.25) depend on the variables a and b. Hence, the loop of lines 3 to 6 of Algorithm 1 must be executed two times for $i = 1$ and $i = 2$.

The first sweep of the loop in Algorithm 1 generates and uses the set $ST(\mathbf{v})$ with the exchanged columns $v_0 \Leftrightarrow v_1$ and $v_2 \Leftrightarrow v_3$. The indices of these pairs of variables can be taken from the first two rows in column $i = 1$ of Tab. 3.9. The first elements of these pairs are associated with the variable tuple 6 in lines 15 and 16 of the PRP listing in Fig. 3.3. In the same order the second elements of these pairs are associated with the variable tuple 7 in lines 17 and 18 of the PRP. The XBOOLE operation cco changes the columns in $XBO[5]$ controlled by the variable tuples $XBO[6]$ and $XBO[7]$ into the final XBOOLE object $XBO[8]$ (3.89) which describes the same set as $ST(\mathbf{v})$ (3.79).

$$XBO[8] = ST(\mathbf{v}) = \begin{array}{cccc} v_0 & v_1 & v_2 & v_3 \\ \hline 0 & 1 & 1 & 1 \\ 1 & 0 & 0 & 0 \\ 1 & 1 & 1 & 0 \\ 0 & 1 & 1 & 0 \\ 1 & 0 & 0 & 1 \\ 0 & 0 & 0 & 1 \\ 0 & 0 & 1 & 0 \\ 1 & 1 & 0 & 1 \\ \hline \end{array} \qquad (3.89)$$

The intersection of $XBO[5]$ (3.88) and $XBO[8]$ (3.89) in line 20 of the PRP calculates the XBOOLE object $XBO[9]$ (3.90) which contains the same six vectors $\mathbf{v}$ as $S(\mathbf{v})$ (3.80).

$$XBO[9] = S(\mathbf{v}) = \begin{array}{cccc} v_0 & v_1 & v_2 & v_3 \\ \hline 1 & 1 & 0 & 1 \\ 1 & 0 & 0 & 1 \\ 0 & 1 & 1 & 0 \\ 0 & 0 & 1 & 0 \\ 0 & 0 & 0 & 1 \\ 1 & 1 & 1 & 0 \\ \hline \end{array} \qquad (3.90)$$

The second sweep of the loop in Algorithm 1 requires the set $ST(\mathbf{v})$ with the exchanged columns $v_0 \Leftrightarrow v_2$ and $v_1 \Leftrightarrow v_3$. The indices of these pairs of variables can be taken from the first two rows in column $i = 2$ of Tab. 3.9. The first elements of these pairs are assigned to the variable tuple $XBO[10]$ in lines 21 and 22, and the second elements of these pairs are associated with the variable tuple $XBO[11]$ in lines 23 and 24 in the same order. The XBOOLE operation cco changes the columns in $XBO[9]$ controlled by the variable tuple $XBO[10]$ and $XBO[11]$ into the XBOOLE object $XBO[12]$ (3.91) which describes the same set as $ST(\mathbf{v})$ (3.81).

$$XBO[12] = ST(\mathbf{v}) = \begin{array}{cccc} v_0 & v_1 & v_2 & v_3 \\ \hline 0 & 1 & 1 & 1 \\ 0 & 1 & 1 & 0 \\ 1 & 0 & 0 & 1 \\ 1 & 0 & 0 & 0 \\ 0 & 1 & 0 & 0 \\ 1 & 0 & 1 & 1 \\ \hline \end{array} \qquad (3.91)$$

The intersection of $XBO[9]$ (3.90) and $XBO[12]$ (3.91) in line 26 of the PRP calculates the final XBOOLE object $XBO[13]$ (3.92) which contains the same two vectors $\mathbf{v}$ as $S(\mathbf{v})$ (3.82). Hence, these two vectors describe based on the normal form (3.83) the set of solution functions (3.84).

$$XBO[13] = S(\mathbf{v}) = \begin{array}{cccc} v_0 & v_1 & v_2 & v_3 \\ \hline 1 & 0 & 0 & 1 \\ 0 & 1 & 1 & 0 \\ \hline \end{array} \qquad (3.92)$$

As a second example, we solve a more complex Boolean differential equation which was specified in Steinbach and Posthoff [2011]. The BDE (3.93) describes the set of most complex bent functions of four variables.

$$\begin{aligned}
&\left(\frac{\partial f(\mathbf{x})}{\partial x_1} \oplus \frac{\partial f(\mathbf{x})}{\partial x_2} \oplus \frac{\partial f(\mathbf{x})}{\partial (x_1, x_2)}\right) \wedge \left(\frac{\partial f(\mathbf{x})}{\partial x_1} \oplus \frac{\partial f(\mathbf{x})}{\partial x_3} \oplus \frac{\partial f(\mathbf{x})}{\partial (x_1, x_3)}\right) \wedge \\
&\left(\frac{\partial f(\mathbf{x})}{\partial x_1} \oplus \frac{\partial f(\mathbf{x})}{\partial x_4} \oplus \frac{\partial f(\mathbf{x})}{\partial (x_1, x_4)}\right) \wedge \left(\frac{\partial f(\mathbf{x})}{\partial x_2} \oplus \frac{\partial f(\mathbf{x})}{\partial x_3} \oplus \frac{\partial f(\mathbf{x})}{\partial (x_2, x_3)}\right) \wedge \\
&\left(\frac{\partial f(\mathbf{x})}{\partial x_2} \oplus \frac{\partial f(\mathbf{x})}{\partial x_4} \oplus \frac{\partial f(\mathbf{x})}{\partial (x_2, x_4)}\right) \wedge \left(\frac{\partial f(\mathbf{x})}{\partial x_3} \oplus \frac{\partial f(\mathbf{x})}{\partial x_4} \oplus \frac{\partial f(\mathbf{x})}{\partial (x_3, x_4)}\right) = 1 .
\end{aligned} \qquad (3.93)$$

Rothaus [1976] introduced bent functions for the first time. Bent functions are Boolean functions that have the largest Hamming distance to all linear Boolean functions. Therefore, bent functions are very useful for cryptographic applications. We have selected the BDE (3.93) not only as an example to solve one more BDE of the type (3.23) but also to explain a modified solution procedure.

The BDE (3.93) depends on all four simple derivatives of $f(\mathbf{x}) = f(x_1, x_2, x_3, x_4)$ and additionally on six vectorial derivatives of the same function. This BDE specifies a certain set of all $2^{2^4} = 65{,}536$ Boolean functions of four variables. This number of Boolean functions already has a size for which a computer must be used; a solution by hand is not possible.

The problem program of Fig. 3.3 can be adapted to solve the BDE (3.93). However, two properties of the used procedure cause an unnecessarily large effort.

The first of these properties is the requirement to solve the system of Boolean equations (3.71), (3.72). Due to the used $n = 4$ variables, this system of equations consists of the simple equation (3.71) and $2^4 - 1 = 15$ equations (3.72) which contain the linear EXOR operation. Hence, the solution of this system of Boolean equations consists of 65,536 Boolean vectors of 32 variables which describe the mapping of all vectors $\mathbf{u} = (u_0, u_1, \ldots, u_{15})$ to all vectors $\mathbf{v} = (v_0, v_1, \ldots, v_{15})$. However, out of all vectors $\mathbf{u}$ only the vectors are needed that solve the associated Boolean equation of the BDE (3.93).

The simple structure of the system of Boolean equations (3.71), (3.72) provides a possibility for the direct mapping of the vectors $\mathbf{u}$ of local solutions of the BDE (3.93) into the set of vectors of function values $\mathbf{v}$ of potential solution functions. The variable u_0 appears in each of the 16 equations. For the constant value $u_0 = 0$ the system of equations (3.71), (3.72) is changed into:

$$v_0 = 0, \tag{3.94}$$
$$v_i = u_i, \quad \text{with} \quad i = 1, 2, \ldots, 2^n - 1, \tag{3.95}$$

and for the constant value $u_0 = 1$ we have:

$$v_0 = 1, \tag{3.96}$$
$$v_i = \overline{u}_i, \quad \text{with} \quad i = 1, 2, \ldots, 2^n - 1, \tag{3.97}$$

The required mapping of the set of local solutions $SLS(\mathbf{u})$ into the set of vectors of function values $S(\mathbf{v})$ of potential solution functions can be simplified to (3.98) utilizing the equations (3.94), ..., (3.97):

$$S(\mathbf{v}) = \overline{v}_0 \wedge \left[\max_{u_0} (\overline{u}_0 \wedge SLS(\mathbf{u})) \right]_{u_i \to v_i} \vee v_0 \wedge \left[\max_{u_0} (u_0 \wedge SLS(\overline{\mathbf{u}})) \right]_{u_i \to v_i}. \tag{3.98}$$

One more property of the procedure used in the PRP of Fig. 3.3 is the use of the sets of variables $\mathbf{u}$ and $\mathbf{v}$. These sets of variables are helpful to explain and understand the solution steps of a BDE (3.22). Both sets contain the same number of Boolean variables. This number is equal to 2^n for a function of n variables in the BDE. The different variables have different meanings. The variables v_i, $i = 0, \ldots, 2^n - 1$, already change the meaning of the stored data during their lifetime. These variables are initialized with function values of potential solution functions and contain at the end of Algorithm 1 the actual solution functions.

Algorithm 1 needs at each point in time only one set of 2^n variables to store data of a certain meaning. Hence, a single set of 2^n variables is sufficient to solve a BDE for a function of n variables. Due to the changing meaning the variables v_i are used to substitute for the variables u_i. With this assumption the Boolean equation associated with the BDE (3.93) is:

$$(v_{01} \oplus v_{02} \oplus v_{03}) \wedge (v_{01} \oplus v_{04} \oplus v_{05}) \wedge$$
$$(v_{01} \oplus v_{08} \oplus v_{09}) \wedge (v_{02} \oplus v_{04} \oplus v_{06}) \wedge$$
$$(v_{02} \oplus v_{08} \oplus v_{10}) \wedge (v_{04} \oplus v_{08} \oplus v_{12}) = 1. \tag{3.99}$$

Due to the exponential complexity of Boolean problems, the bisection of the numbers of Boolean variables is very beneficial for their solution. An additional advantage for solving a BDE

(3.22) is that the mapping Eq. (3.98) can be simplified to (3.100):

$$S(\mathbf{v}) \leftarrow \overline{v}_0 \wedge \left[\max_{v_0} SLS(\mathbf{v}) \right] \vee v_0 \wedge \left[\max_{v_0} SLS(\overline{\mathbf{v}}) \right] . \tag{3.100}$$

Utilizing these two simplifications, the BDE (3.93) can be solved by the PRP shown in Fig. 3.4. After the definition of a Boolean space of 32 variables all required 16 variables $v_{00}, \ldots, v_{15}$ are assigned to this space in lines 2 to 6 of this PRP. Next, the associated Boolean equations (3.99) of the BDE (3.93) are solved in lines 7 to 13 of the PRP in Fig. 3.4 and stored as XBOOLE object 1 (3.101). This set $SLS(\mathbf{v})$ (3.101) contains 16 Boolean vectors.

	v_{01}	v_{02}	v_{03}	v_{04}	v_{05}	v_{06}	v_{08}	v_{09}	v_{10}	v_{12}	
	1	1	1	1	1	1	1	1	1	1	
	1	1	1	1	1	1	0	0	0	0	
	1	0	0	1	1	0	1	1	0	1	
	0	1	0	1	0	1	1	0	1	1	
	0	0	1	0	1	1	0	1	1	1	
	1	0	0	0	0	1	0	0	1	1	
	0	1	0	0	1	0	0	1	0	1	
$XBO[1] = SLS(\mathbf{v}) =$	0	0	1	1	0	0	1	0	0	1	(3.101)
	1	1	1	0	0	0	0	0	0	1	
	0	0	1	0	1	1	1	0	0	0	
	1	1	1	0	0	0	1	1	1	0	
	1	0	0	0	0	1	1	1	0	0	
	0	1	0	1	0	1	0	1	0	0	
	0	0	1	1	0	0	0	1	1	0	
	0	1	0	0	1	0	1	0	1	0	
	1	0	0	1	1	0	0	0	1	0	

Due to the missing six variables (v_{00}, v_{07}, v_{11}, v_{13}, v_{14}, v_{15}) the XBOOLE object $XBO[1]$ describes the set $SLS(\mathbf{v})$ of $16 * 2^6 = 1{,}024$ local solutions.

A particularity of the BDE (3.93) is that the function $f(\mathbf{x})$ does not occur directly without a derivative. Therefore, the variable v_{00} does not appear in the associated Boolean equation (3.99) so that it is not necessary to calculate the maximum operations with regard to v_0 in (3.100).

The transformation (3.100) happens in lines 14 to 20 of the PRP. The term $\overline{v}_{00} \wedge SLS(\mathbf{v})$ is built in lines 14 to 16 and stored as XBOOLE object 3. The second term $v_{00} \wedge SLS(\overline{\mathbf{v}})$ of (3.100) is calculated in lines 17 to 19 of the PRP. The complement in line 17 changes the XBOOLE object 2 from $\overline{v}_{00}$ to v_{00}. The XBOOLE operation cel changes all elements $0 \Rightarrow 1$ and $1 \Rightarrow 0$ in the set of local solutions $SLS(\mathbf{v})$ so that $SLS(\overline{\mathbf{v}})$ is created and stored as XBOOLE object 4. The union of the XBOOLE objects 3 and 4 in line 20 summarizes both terms of (3.100) and stores the set of potential solution vectors as XBOOLE object 5.

The remaining lines of the PRP in Fig. 3.4 are the four required sweeps of the loop of Algorithm 1. The indices of the variables which describe the columns to exchange are taken from

```
1    space 32 1                      27   cco 5 6 7 8
2    avar 1                          28   isc 5 8 9
3    v00 v01 v02 v03                 29   vtin 1 10
4    v04 v05 v06 v07                 30   v00 v01 v04 v05
5    v08 v09 v10 v11                 31   v08 v09 v12 v13.
6    v12 v13 v14 v15.                32   vtin 1 11
7    sbe 1 1                         33   v02 v03 v06 v07
8    (v01#v02#v03)&                  34   v10 v11 v14 v15.
9    (v01#v04#v05)&                  35   cco 9 10 11 12
10   (v01#v08#v09)&                  36   isc 9 12 13
11   (v02#v04#v06)&                  37   vtin 1 14
12   (v02#v08#v10)&                  38   v00 v01 v02 v03
13   (v04#v08#v12)=1.                39   v08 v09 v10 v11.
14   sbe 1 2                         40   vtin 1 15
15   v00=0.                          41   v04 v05 v06 v07
16   isc 1 2 3                       42   v12 v13 v14 v15.
17   cpl 2 2                         43   cco 12 14 15 16
18   cel 1 1 4 /01 /10               44   isc 13 16 17
19   isc 2 4 4                       45   vtin 1 18
20   uni 3 4 5                       46   v00 v01 v02 v03
21   vtin 1 6                        47   v04 v05 v06 v07.
22   v00 v02 v04 v06                 48   vtin 1 19
23   v08 v10 v12 v14.                49   v08 v09 v10 v11
24   vtin 1 7                        50   v12 v13 v14 v15.
25   v01 v03 v05 v07                 51   cco 17 18 19 20
26   v09 v11 v13 v15.                52   isc 17 20 21
```

Figure 3.4: Listing of the PRP to solve the BDE (3.93).

the first eight lines of Tab. 3.9. The left variables from column $i = 1$ of Tab. 3.9 are assigned to the variable tuple 6, and the right variables of this column are assigned in the same order to the variable tuple 7. These two variable tuples control the exchange of columns of $S(\mathbf{v})$ stored as XBOOLE object 5 into $ST(\mathbf{v})$ stored as XBOOLE object 8. The intersection of these two XBOOLE objects in line 28 of the PRP calculates the restricted set $S(\mathbf{v})$ and stores it as XBOOLE object 9.

$$XBO[21] = S(\mathbf{v}) =$$

v_{00}	v_{01}	v_{02}	v_{03}	v_{04}	v_{05}	v_{06}	v_{07}	v_{08}	v_{09}	v_{10}	v_{11}	v_{12}	v_{13}	v_{14}	v_{15}	
0	1	1	1	1	1	1	0	1	1	1	0	1	0	0	0	1
0	1	1	1	1	1	1	0	0	0	0	1	0	1	1	1	2
0	1	0	0	1	1	0	1	1	1	0	1	1	0	1	1	3
0	0	1	0	1	0	1	1	1	0	1	1	1	1	0	1	4
0	0	0	1	0	1	1	1	0	1	1	1	1	1	1	0	5
0	1	0	0	0	0	1	0	0	0	1	0	1	0	1	1	6
0	0	1	0	0	1	0	0	0	1	0	0	1	1	0	1	7
0	0	0	1	1	0	0	0	1	0	0	0	1	1	1	0	8
0	1	1	1	0	0	0	1	0	0	0	1	1	0	0	0	9
0	0	0	1	0	1	1	1	1	0	0	0	0	0	0	1	10
0	1	1	1	0	0	0	1	1	1	1	0	0	1	1	1	11
0	1	0	0	0	0	1	0	1	1	0	1	0	1	0	0	12
0	0	1	0	1	0	1	1	0	1	0	0	0	0	1	0	13
0	0	0	1	1	0	0	0	0	1	1	1	0	0	0	1	14
0	0	1	0	0	1	0	0	1	0	1	1	0	0	1	0	15
0	1	0	0	1	1	0	1	0	0	1	0	0	1	0	0	16
1	0	0	0	0	0	1	0	0	0	1	0	1	1	1	1	17
1	0	0	0	0	0	0	1	1	1	1	0	1	0	0	0	18
1	0	1	1	0	0	1	0	0	0	1	0	0	1	0	0	19
1	1	0	1	0	1	0	0	0	1	0	0	0	0	1	0	20
1	1	1	0	1	0	0	0	1	0	0	0	0	0	0	1	21
1	0	1	1	1	1	0	1	1	1	0	1	0	1	0	0	22
1	1	0	1	1	0	1	1	1	0	1	1	0	0	1	0	23
1	1	1	0	0	1	1	1	0	1	1	1	0	0	0	1	24
1	0	0	0	1	1	1	0	1	1	1	0	0	1	1	1	25
1	1	1	0	1	0	0	0	0	1	1	1	1	1	1	0	26
1	0	0	0	1	1	1	0	0	0	0	1	1	0	0	0	27
1	0	1	1	1	1	0	1	0	0	1	0	1	0	1	1	28
1	1	0	1	0	1	0	0	1	0	1	1	1	1	0	1	29
1	1	1	0	0	1	1	1	1	0	0	0	1	1	1	0	30
1	1	0	1	1	0	1	1	0	1	0	0	1	1	0	1	31
1	0	1	1	0	0	1	0	1	1	0	1	1	0	1	1	32

$$(3.102)$$

The next three iterative executions of the operations of lines 4 and 5 of Algorithm 1 calculate the final solution (3.102). These operations are realized:

- in lines 29 to 36 of the PRP for $i = 2$,

- in lines 37 to 44 of the PRP for $i = 3$, and

- in lines 45 to 52 of the PRP for $i = 4$.

The indices of the variables v_j that indicate the columns of $S(\mathbf{v})$ which must be exchanged are taken in each iteration from the first eight rows of Tab. 3.9 where the value of i selects the appropriate column.

The actual set of vectors of function values $S(\mathbf{v})$ (3.102) is stored as XBOOLE object 21 in the PRP of Fig. 3.4 and contains 32 solution vectors. The solution functions $f(x_1, x_2, x_3, x_4)$ are defined by these vectors of function values and the adapted normal form of (3.74):

$$f(x_1, x_2, x_3, x_4) =$$
$$\overline{x}_1\,\overline{x}_2\,\overline{x}_3\,\overline{x}_4\,v_{00} \oplus x_1\,\overline{x}_2\,\overline{x}_3\,\overline{x}_4\,v_{01} \oplus \overline{x}_1\,x_2\,\overline{x}_3\,\overline{x}_4\,v_{02} \oplus x_1\,x_2\,\overline{x}_3\,\overline{x}_4\,v_{03} \oplus$$
$$\overline{x}_1\,\overline{x}_2\,x_3\,\overline{x}_4\,v_{04} \oplus x_1\,\overline{x}_2\,x_3\,\overline{x}_4\,v_{05} \oplus \overline{x}_1\,x_2\,x_3\,\overline{x}_4\,v_{06} \oplus x_1\,x_2\,x_3\,\overline{x}_4\,v_{07} \oplus$$
$$\overline{x}_1\,\overline{x}_2\,\overline{x}_3\,x_4\,v_{08} \oplus x_1\,\overline{x}_2\,\overline{x}_3\,x_4\,v_{09} \oplus \overline{x}_1\,x_2\,\overline{x}_3\,x_4\,v_{10} \oplus x_1\,x_2\,\overline{x}_3\,x_4\,v_{11} \oplus$$
$$\overline{x}_1\,\overline{x}_2\,x_3\,x_4\,v_{12} \oplus x_1\,\overline{x}_2\,x_3\,x_4\,v_{13} \oplus \overline{x}_1\,x_2\,x_3\,x_4\,v_{14} \oplus x_1\,x_2\,x_3\,x_4\,v_{15}\,. \tag{3.103}$$

The solution of the BDE consists of two function classes. The first of these classes contains 16 functions with 10 function values 1. The substitution of the values of row number 5, which is colored in (3.102), into the normal form (3.103) can be used as a representative of the first solution class:

$$f(x_1, x_2, x_3, x_4) = x_1\,x_2 \oplus x_1\,x_3 \oplus x_1\,x_4 \oplus x_2\,x_3 \oplus x_2\,x_4 \oplus x_3\,x_4\,. \tag{3.104}$$

The other 15 functions of this class have the same structure as (3.104) and are built as defined in (3.51) by all combinations of negations of the four variables.

The second solution class of the BDE (3.93) contains also 16 functions, but each of these functions has only six function values 1. The substitution of the function values of row number 21, which is also colored in (3.102), into the normal form (3.103) can be used as representative of the second solution class:

$$f(x_1, x_2, x_3, x_4) = 1 \oplus x_1\,x_2 \oplus x_1\,x_3 \oplus x_1\,x_4 \oplus x_2\,x_3 \oplus x_2\,x_4 \oplus x_3\,x_4\,. \tag{3.105}$$

The comparison of both colored rows in (3.102) shows that the representative solution functions of both classes are complements of each other. Hence, (3.105) extends (3.104) by an EXOR with the constant 1. The other 15 functions of the second solution class have the same structure as (3.105) and are also built as defined in (3.51) by all combinations of negations of the four variables.

3.5   BOOLEAN DIFFERENTIAL EQUATIONS FOR EACH SET OF SOLUTION FUNCTIONS

3.5.1   SEPARATION OF FUNCTIONS

Boolean differential equations of the type (3.22) are restricted to derivatives with regard to all directions of change. We learned in Sec. 3.4 that the solution of such BDEs are sets of Boolean functions of certain function classes. That means, either all functions of a function class defined by the relation (3.58) belong to the solution set or not any of these Boolean functions.

This property restricts the application of BDEs because not all sets of Boolean functions can be expressed by a BDE of the type (3.22). The extension of the BDE (3.22) by the Boolean variables $\mathbf{x} = (x_1, x_2, \ldots, x_n)$ of the function $f(\mathbf{x})$ eliminates the restriction to function classes as a result of a BDE. Therefore, we generalize the explored BDEs in this section such that the BDE can contain simple and vectorial derivatives of the same function, this function $f(\mathbf{x})$ itself, and the Boolean variables $\mathbf{x}$ in expressions D_1 and D_2 of both sides of the BDE:

$$D_1\left(f(\mathbf{x}), \frac{\partial f(\mathbf{x})}{\partial x_1}, \frac{\partial f(\mathbf{x})}{\partial x_2}, \dots, \frac{\partial f(\mathbf{x})}{\partial \mathbf{x}}, \mathbf{x}\right) = D_2\left(f(\mathbf{x}), \frac{\partial f(\mathbf{x})}{\partial x_1}, \frac{\partial f(\mathbf{x})}{\partial x_2}, \dots, \frac{\partial f(\mathbf{x})}{\partial \mathbf{x}}, \mathbf{x}\right). \quad (3.106)$$

A BDE of the type (3.106) can be solved in a trivial but expensive manner. Each of the 2^{2^n} Boolean functions $f(x_1, x_2, \dots, x_n)$ can be substituted into the BDE (3.106). Results of all calculated derivatives are Boolean expressions. Hence, the BDE (3.106) is changed for the selected Boolean function $f(x_1, x_2, \dots, x_n)$ into a Boolean equation that must be satisfied for all $\mathbf{x} = \mathbf{c}$ of the Boolean space B^n.

The following example demonstrates this approach for the simple BDE:

$$a \wedge f(a, b) \wedge \frac{\partial f(a, b)}{\partial a} \vee \overline{b} = \overline{b} \quad (3.107)$$

which describes the set of Boolean functions which are monotonously rising in the direction of a for $b = 1$.

First, we verify whether the function

$$f(a, b) = a \oplus b \quad (3.108)$$

satisfies the BDE (3.107). Using

$$\frac{\partial f(a, b)}{\partial a} = \frac{\partial (a \oplus b)}{\partial a} = 0 \oplus b \oplus 1 \oplus b = 1 \quad (3.109)$$

the BDE (3.107) is simplified into the Boolean equation:

$$a \wedge (a \oplus b) \wedge 1 \vee \overline{b} = \overline{b} \quad (3.110)$$

which can be simplified to:

$$a \wedge (\overline{a} b \vee a \overline{b}) \wedge 1 \vee \overline{b} = \overline{b}$$
$$a \overline{b} \vee \overline{b} = \overline{b}$$
$$\overline{b} \equiv \overline{b} . \quad (3.111)$$

The Boolean equation (3.111) is satisfied for all values of b. Hence, the function (3.108) is a solution function of the BDE (3.107).

The Boolean function:

$$f(a, b) = \overline{a} \oplus b \quad (3.112)$$

belongs to the same function class as the function (3.108). We verify as counterexample whether the function (3.112) satisfies the BDE (3.107). Using

$$\frac{\partial f(a, b)}{\partial a} = \frac{\partial (\overline{a} \oplus b)}{\partial a} = 1 \oplus b \oplus 0 \oplus b = 1 \quad (3.113)$$

the BDE (3.107) is simplified into the Boolean equation:

$$a \wedge (\bar{a} \oplus b) \wedge 1 \vee \bar{b} \stackrel{?}{=} \bar{b} \tag{3.114}$$

which can be simplified to:

$$a \wedge (a\,b \vee \bar{a}\,\bar{b}) \wedge 1 \vee \bar{b} \stackrel{?}{=} \bar{b}$$
$$a\,b \vee \bar{b} \stackrel{?}{=} \bar{b}$$
$$a \vee \bar{b} \stackrel{?}{=} \bar{b} \,. \tag{3.115}$$

The Boolean equation (3.115) is not satisfied for $(a, b) = (1, 1)$. Hence, the function (3.112) is not a solution function of the BDE (3.107). This example verifies that one function, but not all functions of a class (3.58) of Boolean functions, belongs to the solution set of the BDE (3.107) of the type (3.106). A drawback of this approach is that all 2^{2^n} Boolean functions of B^n must be checked.

A more efficient solution algorithm for BDEs of the type (3.106) was developed in the Ph.D. thesis of Steinbach [1981]. This algorithm generalizes the approach of separation of function classes to the separation of arbitrary function sets. The extension of the BDE (3.22) to the BDE (3.106) by the Boolean variables $\mathbf{x}$ requires an extension of the definition of the term *local solution* an $\nabla f (\mathbf{x})$.

Definition 3.6   Let $f(\mathbf{x})$ be a solution function of a BDE (3.106). Then

$$LS = \left[f(\mathbf{x}), \frac{\partial f(\mathbf{x})}{\partial x_1}, \frac{\partial f(\mathbf{x})}{\partial x_2}, \dots, \frac{\partial f(\mathbf{x})}{\partial \mathbf{x}}, \mathbf{x} \right]_{\mathbf{x}=\mathbf{c}} \tag{3.116}$$

is a *local solution* for $\mathbf{x} = \mathbf{c}$, and

$$(\nabla f(\mathbf{x}), \mathbf{x}) = \left(f(\mathbf{x}), \frac{\partial f(\mathbf{x})}{\partial x_1}, \frac{\partial f(\mathbf{x})}{\partial x_2}, \dots, \frac{\partial f(\mathbf{x})}{\partial \mathbf{x}}, \mathbf{x} \right) \tag{3.117}$$

is a short notation for the vector of $f(\mathbf{x})$, all simple and vectorial derivatives of this Boolean function, and the Boolean variables $\mathbf{x} = (x_1, x_2, \dots, x_n)$.

A local solution (3.116) is a vector of $n + 2^n$ Boolean values that determine:

1. one point $\mathbf{x} = \mathbf{c}$ of the Boolean space B^n by n values,

2. the function value $f(\mathbf{c})$, and

3. $2^n - 1$ values of all derivatives with regard to all directions of change of the same function at the same point $\mathbf{x} = \mathbf{c}$.

The information of such a local solution is sufficient to reconstruct one complete potential function $f(\mathbf{x})$, because the local function value and its changes for all directions are known.

The idea of the separation of a function set as solution of a BDE (3.106) is similar to the separation of function classes of a BDE (3.22). A BDE of the type (3.106) contains only the elements of $(\nabla f(\mathbf{x}), \mathbf{x})$ defined by (3.117) either in non-negated or negated form. Hence, these elements of the BDE (3.106) can be modeled by Boolean variables, but Def. 3.5 must be adapted for the BDE (3.106).

Definition 3.7   Based on the Boolean differential equation (3.106) we define the associated Boolean equation

$$D_1(u_0, u_1, ..., u_{2^n-1}, x_1, x_2, \ldots, x_n) = D_2(u_0, u_1, ..., u_{2^n-1}, x_1, x_2, \ldots, x_n) . \qquad (3.118)$$

The index j of the variable u_j indicates the associated element of the BDE (3.106) such that a value 1 in the binary code of the index determines a variable in the change vector of the associated derivative. Using this rule, we map

- the function $f(\mathbf{x})$ to u_0 ,

- the derivative $\dfrac{\partial f(\mathbf{x})}{\partial x_1}$ to u_1 ,

- the derivative $\dfrac{\partial f(\mathbf{x})}{\partial x_2}$ to u_2 ,

- the derivative $\dfrac{\partial f(\mathbf{x})}{\partial(x_1, x_2)}$ to u_3 ,

- ..., and finally

- the derivative $\dfrac{\partial f(\mathbf{x})}{\partial \mathbf{x}}$ to u_{2^n-1} .

The solution of the associated Boolean equation (3.118) is the *set of local solutions SLS*.

Knowing the set of local solutions SLS of the associated Boolean equation (3.118), it remains the task to separate local solutions of actual solution functions of the BDE (3.106) from the other local solutions. Boolean functions which do not appear at least for one point of the Boolean space B^n as local solutions cannot belong to the set of solution functions of the BDE (3.106). However, this is only a necessary condition. A necessary and sufficient condition for a solution function $f(x_1, x_2, \ldots, x_n)$ of a BDE (3.106) is that this function appears in all 2^n points of the Boolean space B^n as a local solution (3.116):

$$\forall \mathbf{c} \in B^n \quad (\nabla f(\mathbf{x}), \mathbf{x}) \mid_{\mathbf{x}=\mathbf{c}} = (\mathbf{u}, \mathbf{x}) \mid_{\mathbf{x}=\mathbf{c}} \in SLS . \qquad (3.119)$$

The correctness of condition (3.119) follows from the property that a solution function has local solutions for each point $\mathbf{x} = \mathbf{c}$ of B^n. Otherwise, the Boolean functions on the left-hand side and on the right-hand side of the Boolean equation, which is generated by the substitution of a function which is not a solution into the BDE (3.106), are not identical functions.

It is the aim of the procedure of function separation to find all functions $f(\mathbf{x})$ which satisfy the condition (3.119). Two subtasks must be solved to achieve this goal:

1. all vectors $(\mathbf{u}, \mathbf{x})$ of local solutions which do not satisfy the condition (3.119) must be excluded from the set SLS, and

2. each subset of 2^n vectors $(\mathbf{u}, \mathbf{x})$ of local solutions that describes an individual solution function $f(\mathbf{x})$ must be mapped to this function.

Due to the variables $\mathbf{x}$ in the set of local solutions $SLS(\mathbf{u}, \mathbf{x})$, it is more difficult to separate all particular solution functions of the BDE (3.106) from $SLS(\mathbf{u}, \mathbf{x})$. For that reason we explore first an example and generalize then the solution procedure.

Fig. 3.5 shows one solution function of the BDE (3.107). For all four points of the Boolean space the local solutions $(\nabla f(a, b), a, b)$ are given. These local solutions can be verified directly by means of the depicted function in the Boolean space B^2 and the values of the derivatives which are indicated by different styles of the connection lines. A comma in such a vector separates values of $\nabla f(a, b)$ from the values of (a, b). The comparison of the four vectors $(\nabla f(a, b), a, b)$ shows that not only the values of (a, b) but also all Boolean vectors of $\nabla f(a, b)$ are different.

Next, we verify that the four local solutions belong to the set of local solutions $SLS(\mathbf{u}, \mathbf{x})$ of the associated Boolean equation:

$$a \wedge u_0 \wedge u_1 \vee \overline{b} = \overline{b} \tag{3.120}$$

of the BDE (3.107):

$$
\begin{aligned}
(u_0\, u_1\, u_2\, u_3, a\, b) &= (1100, 00): &&0 \wedge 1 \wedge 1 \vee \overline{0} = \overline{0} \;\Rightarrow\; 1 = 1\,, \\
(u_0\, u_1\, u_2\, u_3, a\, b) &= (1001, 01): &&0 \wedge 1 \wedge 0 \vee \overline{1} = \overline{1} \;\Rightarrow\; 0 = 0\,, \\
(u_0\, u_1\, u_2\, u_3, a\, b) &= (0111, 10): &&1 \wedge 0 \wedge 1 \vee \overline{0} = \overline{0} \;\Rightarrow\; 1 = 1\,, \\
(u_0\, u_1\, u_2\, u_3, a\, b) &= (1010, 11): &&1 \wedge 1 \wedge 0 \vee \overline{1} = \overline{1} \;\Rightarrow\; 0 = 0\,.
\end{aligned}
\tag{3.121}
$$

All four local solution vectors $(\mathbf{u}, a, b)$ of the function $f(a, b) = \overline{a} \vee b$ belong to the set of local solutions $SLS(\mathbf{u}, a, b)$ of the associated Boolean equation (3.120).

The procedure of the function separation must verify that the four vectors $(\mathbf{u}, a, b)$ shown in Fig. 3.5 are elements of the set of local solutions $SLS(\mathbf{u}, a, b)$. The fact that these vectors $\mathbf{u}$ contain different numbers of values 1 complicates the verification. The reason for this problem is that u_0 describes a function value but the values $u_1, u_2, \ldots, u_{2^n-1}$ describe values of the derivatives with regard to all directions of change in a certain point of the Boolean space. Using the known system of Boolean equations (3.71), and (3.72), the transformation of the set $SLS(\mathbf{u}, a, b)$ into the set $S(\mathbf{v}, a, b)$ can be executed.

This supplementary task realizes the function d2v () (*derivative to value*). After the transformation $S(\mathbf{v}, a, b) \leftarrow \texttt{d2v}(SLS(\mathbf{u}), a, b)$ the set $S(\mathbf{v}, a, b)$ still contains local solutions. However, this set can be restricted in further steps to the set $S(\mathbf{v})$ of the actual solution functions of the BDE (3.106). For that reason the name S (*solution*) is already used. Fig. 3.5 shows for each local solution vector $S(\mathbf{u}, a, b)$ the resulting vector $S(\mathbf{v}, a, b)$ built by the function d2v ().

Each $\mathbf{v}$-part in these vectors describes the same function with three function values 1. However, the meaning of the values v_j, $j = 0, \ldots, 2^n - 1$ is different and depends on the values of the

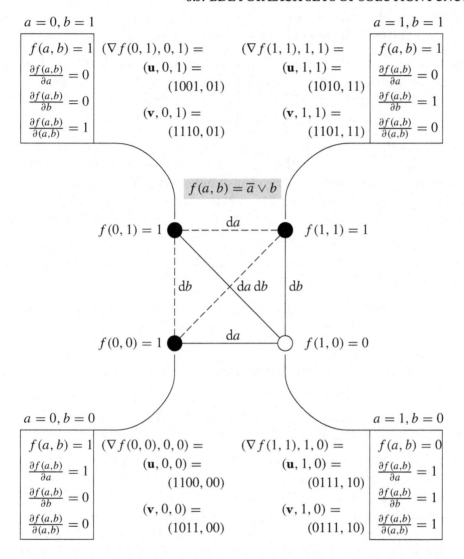

$a = 0, b = 1$

$f(a, b) = 1$

$\frac{\partial f(a,b)}{\partial a} = 0$

$\frac{\partial f(a,b)}{\partial b} = 0$

$\frac{\partial f(a,b)}{\partial (a,b)} = 1$

$(\nabla f(0, 1), 0, 1) =$
$(\mathbf{u}, 0, 1) =$
$(1001, 01)$
$(\mathbf{v}, 0, 1) =$
$(1110, 01)$

$(\nabla f(1, 1), 1, 1) =$
$(\mathbf{u}, 1, 1) =$
$(1010, 11)$
$(\mathbf{v}, 1, 1) =$
$(1101, 11)$

$a = 1, b = 1$

$f(a, b) = 1$

$\frac{\partial f(a,b)}{\partial a} = 0$

$\frac{\partial f(a,b)}{\partial b} = 1$

$\frac{\partial f(a,b)}{\partial (a,b)} = 0$

$f(a, b) = \overline{a} \vee b$

$f(0, 1) = 1$ ---- $\mathrm{d}a$ ---- $f(1, 1) = 1$

$\mathrm{d}b$ $\mathrm{d}a\,\mathrm{d}b$ $\mathrm{d}b$

$f(0, 0) = 1$ ---- $\mathrm{d}a$ ---- $f(1, 0) = 0$

$a = 0, b = 0$

$f(a, b) = 1$

$\frac{\partial f(a,b)}{\partial a} = 1$

$\frac{\partial f(a,b)}{\partial b} = 0$

$\frac{\partial f(a,b)}{\partial (a,b)} = 0$

$(\nabla f(0, 0), 0, 0) =$
$(\mathbf{u}, 0, 0) =$
$(1100, 00)$
$(\mathbf{v}, 0, 0) =$
$(1011, 00)$

$(\nabla f(1, 1), 1, 0) =$
$(\mathbf{u}, 1, 0) =$
$(0111, 10)$
$(\mathbf{v}, 1, 0) =$
$(0111, 10)$

$a = 1, b = 0$

$f(a, b) = 0$

$\frac{\partial f(a,b)}{\partial a} = 1$

$\frac{\partial f(a,b)}{\partial b} = 1$

$\frac{\partial f(a,b)}{\partial (a,b)} = 1$

Figure 3.5: Local solutions of the solution function $f(a, b) = \overline{a} \vee b$ of the BDE (3.107).

associated variables x_i, $i = 1, \ldots, n$. For that reason, the direct mapping of the vectors $\mathbf{v}$ to the function $f(a, b)$ requires $2^2 = 4$ different normal forms:

$$a = 0, b = 0: \quad f(a, b) = \overline{a}\,\overline{b}\,v_0 \oplus a\,\overline{b}\,v_1 \oplus \overline{a}\,b\,v_2 \oplus a\,b\,v_3 , \tag{3.122}$$

$$a = 0, b = 1: \quad f(a, b) = \overline{a}\,\overline{b}\,v_2 \oplus a\,\overline{b}\,v_3 \oplus \overline{a}\,b\,v_0 \oplus a\,b\,v_1 , \tag{3.123}$$

$$a = 1, b = 0: \quad f(a, b) = \overline{a}\,\overline{b}\,v_1 \oplus a\,\overline{b}\,v_0 \oplus \overline{a}\,b\,v_3 \oplus a\,b\,v_2 , \tag{3.124}$$

$$a = 1, b = 1: \quad f(a, b) = \overline{a}\,\overline{b}\,v_3 \oplus a\,\overline{b}\,v_2 \oplus \overline{a}\,b\,v_1 \oplus a\,b\,v_0 . \tag{3.125}$$

The substitution of the $\mathbf{v}$-part of the vectors $(\mathbf{v}, a, b)$ of Fig. 3.5 into the appropriate normal form (3.122), ..., (3.125) results in the same solution function for all four local solutions:

$$(\mathbf{v}, 00) = (1011, 00): \quad f(a, b) = \bar{a}\bar{b} \cdot 1 \oplus a\bar{b} \cdot 0 \oplus \bar{a}b \cdot 1 \oplus ab \cdot 1 = \bar{a} \vee b, \quad (3.126)$$
$$(\mathbf{v}, 01) = (1110, 01): \quad f(a, b) = \bar{a}\bar{b} \cdot 1 \oplus a\bar{b} \cdot 0 \oplus \bar{a}b \cdot 1 \oplus ab \cdot 1 = \bar{a} \vee b, \quad (3.127)$$
$$(\mathbf{v}, 10) = (0111, 10): \quad f(a, b) = \bar{a}\bar{b} \cdot 1 \oplus a\bar{b} \cdot 0 \oplus \bar{a}b \cdot 1 \oplus ab \cdot 1 = \bar{a} \vee b, \quad (3.128)$$
$$(\mathbf{v}, 11) = (1101, 11): \quad f(a, b) = \bar{a}\bar{b} \cdot 1 \oplus a\bar{b} \cdot 0 \oplus \bar{a}b \cdot 1 \oplus ab \cdot 1 = \bar{a} \vee b. \quad (3.129)$$

The found identical function of (3.126), ..., (3.129) separates the function $f(a, b) = \bar{a} \vee b$ as an actual solution function of the BDE (3.107) out of the transformed set $S(\mathbf{v}, a, b)$. This way to separate the solution function requires an inadmissible effort. However, these explicit steps to separate an actual solution function from the set $S(\mathbf{v}, a, b)$ reveals the details for an efficient approach. The following enumeration uses the lexicographical order of the assignment $(a, b) \Rightarrow (x_1, x_2)$ for the generalization to the normal form (3.74):

1. the normal form (3.122) for the local solution of the point $a = 0$, $b = 0$ has the same mapping of the values v_j as used for the separation of classes (3.74),

2. the normal form (3.122) is reached for the local solution of the point $a = 1$, $b = 0$ when the values of the pairs of variables $v_0 \leftrightarrow v_1$ and $v_2 \leftrightarrow v_3$ are exchanged,

3. the same exchange of the values of the pairs of variables $v_0 \leftrightarrow v_1$ and $v_2 \leftrightarrow v_3$ maps the normal form for the local solution of the point $a = 1$, $b = 1$ to the normal form for the local solution of the point $a = 0$, $b = 1$, and

4. the normal form (3.122) is reached for the local solution of the point $a = 0$, $b = 1$ when the values of the pairs of variables $v_0 \leftrightarrow v_2$ and $v_1 \leftrightarrow v_3$ are exchanged.

These pairs of values v_j are the same as defined by (3.73) or enumerated in Tab. 3.9. Hence, the supplementary function epv() (*exchange pairs of values*) as introduced for the separation of function classes can be adapted for the separation of functions. The adapted function epv() exchanges function values $\mathbf{v}$ of $S(\mathbf{v}, \mathbf{x})$ using formula (3.73) with respect to a given index i and returns the set $ST(\mathbf{v}, \mathbf{x})$. The set $ST(\mathbf{v}, \mathbf{x})$ can be created for each variable x_i by $ST(\mathbf{v}, \mathbf{x}) \leftarrow \text{epv}(S(\mathbf{v}, \mathbf{x}), i)$. As a conclusion of the previous enumeration the exchange of pairs of values v_j must only be executed for a local solution of the point with $x_i = 1$.

The separation of function of a set $S(\mathbf{v}, \mathbf{x})$ is achieved when the formula:

$$S(\mathbf{v}, \mathbf{x}) = \max_{x_i} [\bar{x}_i \wedge S(\mathbf{v}, \mathbf{x})] \cap \text{epv}\left(\max_{x_i} [x_i \wedge S(\mathbf{v}, \mathbf{x})], i \right) \quad (3.130)$$

is executed iteratively for all variables x_i that appear in the BDE (3.106). The result set $S(\mathbf{v}, \mathbf{x})$ of (3.130) does not depend on the variable x_i due to the maximum operation with regard to x_i.

Fig. 3.6 shows as an example the required transformation steps to separate functions $f(x_1, x_2, x_3)$ for the positive cofactors $\max_{x_i} [x_i \wedge S(\mathbf{v}, \mathbf{x})]$ with regard to $i = 1, 2, 3$. All local

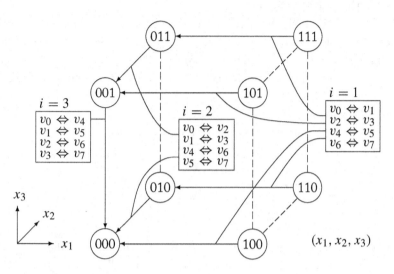

Figure 3.6: Transformation for the separation of functions in the Boolean space B^3.

solutions $S(\mathbf{v}, x_1 = 1, x_2, x_3)$ are mapped in the first separation step to the subspace $x_1 = 0$ and the intersection of (3.130) calculates local solutions $S(\mathbf{v}, x_2, x_3)$. The remaining local solutions $S(\mathbf{v}, x_2 = 1, x_3)$ are mapped in the second separation step to the subspace $x_1 = 0$, $x_2 = 0$ and the intersection of (3.130) calculates local solutions $S(\mathbf{v}, x_3)$. Finally, the local solutions $S(\mathbf{v}, x_3 = 1)$ are mapped in the third separation step to the subspace $x_1 = 0$, $x_2 = 0$, $x_3 = 0$ and the intersection of (3.130) calculates the actual solution functions $S(\mathbf{v})$. Any missing dedicated local solution of a potential function $f_p(x_1, x_2, x_3)$ excludes this function from the set of actual solution functions $S(\mathbf{v})$.

After the iteration with regard to all variables x_i the set $S(\mathbf{v}, \mathbf{x})$ is reduced to the actual solutions $S(\mathbf{v})$ of the BDE (3.106). Each vector $\mathbf{v}$ of the solution set $S(\mathbf{v})$ describes a solution function of the BDE (3.106) by means of the normal form (3.74).

Algorithm 2 shows all steps of the explained solution procedure for BDEs of type (3.106). The BDE to be solved is mapped onto an associated Boolean equation and solved with regard to the set of local solutions $SLS(\mathbf{u}, \mathbf{x})$ in line 1. This set of local information about the function and its derivatives is transformed into the set $S(\mathbf{v}, \mathbf{x})$ of vectors of function values of potential solution functions in line 2. The *separation of functions* which solve the BDE (3.106) is realized in the loop of lines 3 to 8. For the separation of functions the given set $S(\mathbf{v}, \mathbf{x})$ is factorized with regard to x_i in lines 4 and 5 to the subsets $S_0(\mathbf{v}, \mathbf{x} \setminus (x_1, \ldots, x_i))$, $S_1(\mathbf{v}, \mathbf{x} \setminus (x_1, \ldots, x_i))$ which do not depend on the variable x_i. In this way all variables x_i are eliminated from $S(\mathbf{v}, \mathbf{x})$ after all n iterations. The exchange of values in the columns $\mathbf{v}$ by the function epv() is executed in line 6 only for the set $S_1(\mathbf{v}, \mathbf{x} \setminus (x_1, \ldots, x_i))$ into the set $ST_1(\mathbf{v}, \mathbf{x} \setminus (x_1, \ldots, x_i))$. Algorithm 2 needs only n intersections

---

Algorithm 2 Separation of functions

Require: BDE (3.106) in which the function $f(\mathbf{x})$ depends on n variables
Ensure: set S of Boolean vectors $\mathbf{v} = (v_0, v_1, \ldots, v_{2^n-1})$ that describe substituted in (3.74) the set
   of all solution functions of the BDE (3.106)
1: $SLS(\mathbf{u}, \mathbf{x}) \leftarrow$ solution of the Boolean equation (3.118) associated with BDE (3.106)
2: $S(\mathbf{v}, \mathbf{x}) \leftarrow \mathrm{d2v}(SLS(\mathbf{u}, \mathbf{x}))$
3: **for** $i \leftarrow 1$ to n **do**
4:    $S_0(\mathbf{v}, \mathbf{x} \setminus (x_1, \ldots, x_i)) \leftarrow \max_{x_i} [\overline{x}_i \wedge S(\mathbf{v}, x_i, \ldots, x_n)]$
5:    $S_1(\mathbf{v}, \mathbf{x} \setminus (x_1, \ldots, x_i)) \leftarrow \max_{x_i} [x_i \wedge S(\mathbf{v}, x_i, \ldots, x_n)]$
6:    $ST_1(\mathbf{v}, \mathbf{x} \setminus (x_1, \ldots, x_i)) \leftarrow \mathrm{epv}(S_1(\mathbf{v}, \mathbf{x} \setminus (x_1, \ldots, x_i)), i)$
7:    $S(\mathbf{v}, \mathbf{x} \setminus (x_1, \ldots, x_i)) \leftarrow S_0(\mathbf{v}, \mathbf{x} \setminus (x_1, \ldots, x_i)) \cap ST_1(\mathbf{v}, \mathbf{x} \setminus (x_1, \ldots, x_i))$
8: **end for**

---

of sets $S_0(\mathbf{v}, \mathbf{x} \setminus (x_1, \ldots, x_i))$ and $ST_1(\mathbf{v}, \mathbf{x} \setminus (x_1, \ldots, x_i))$ in line 7 for the separation of all solution
functions of a BDE (3.106) out of sets of local solutions for all 2^n points of the Boolean space.

As an example we solve the BDE (3.107) by means of Algorithm 2. (3.120) is the Boolean
equation associated with the BDE (3.107) and has the solution set (3.131).

$$SLS(\mathbf{u}, a, b) = \begin{array}{cccc} u_0 & u_1 & a & b \\ \hline - & - & - & 0 \\ - & - & 0 & 1 \\ - & 0 & 1 & 1 \\ 0 & 1 & 1 & 1 \end{array} \tag{3.131}$$

The result of the transformation of (3.131) into the set $S(\mathbf{v}, a, b)$ based on (3.71) and (3.72) is:

$$S(\mathbf{v}, a, b) = \begin{array}{cccc} v_0 & v_1 & a & b \\ \hline - & - & - & 0 \\ - & - & 0 & 1 \\ - & 1 & 1 & 1 \\ 0 & 0 & 1 & 1 \end{array} . \tag{3.132}$$

The negative cofactor $S_0(\mathbf{v}, a = 0, b)$ and the positive cofactor $S_1(\mathbf{v}, a = 1, b)$ are calculated in
lines 4 and 5 with regard to the variable a.

$$S_0(\mathbf{v}, b) = \begin{array}{ccc} v_0 & v_1 & b \\ \hline - & - & - \end{array} \qquad S_1(\mathbf{v}, b) = \begin{array}{ccc} v_0 & v_1 & b \\ \hline - & - & 0 \\ - & 1 & 1 \\ 0 & 0 & 1 \end{array} \tag{3.133}$$

The ternary representation facilitates very short representations of these sets. Due to the missing
variables v_2 and v_3 in $S_1(\mathbf{v}, b)$ (3.133) the exchange of columns for $i = 1$ is restricted to $v_0 \leftrightarrow v_1$

in the first execution of line 6 of Algorithm 2.

$$ST_1(\mathbf{v}, b) = \begin{array}{ccc} v_0 & v_1 & b \\ \hline - & - & 0 \\ 1 & - & 1 \\ 0 & 0 & 1 \end{array} \tag{3.134}$$

The new set $S(\mathbf{v}, b) \leftarrow S_0(\mathbf{v}, b) \cap ST_1(\mathbf{v}, b)$ (3.130) is built by an intersection as specified in line 7 of Algorithm 2.

$$S(\mathbf{v}, b) = \begin{array}{ccc} v_0 & v_1 & b \\ \hline - & - & - \end{array} \cap \begin{array}{ccc} v_0 & v_1 & b \\ \hline - & - & 0 \\ 1 & - & 1 \\ 0 & 0 & 1 \end{array} = \begin{array}{ccc} v_0 & v_1 & b \\ \hline - & - & 0 \\ 1 & - & 1 \\ 0 & 0 & 1 \end{array} \tag{3.135}$$

In the second sweep of the loop in Algorithm 2 the negative cofactor $S_0(\mathbf{v}, b = 0)$ and the positive cofactor $S_1(\mathbf{v}, b = 1)$ with regard to the variable b are calculated.

$$S_0(\mathbf{v}) = \begin{array}{cc} v_0 & v_1 \\ \hline - & - \end{array} \qquad S_1(\mathbf{v}) = \begin{array}{cc} v_0 & v_1 \\ \hline 1 & - \\ 0 & 0 \end{array} \tag{3.136}$$

The variable $i = 2$ in this sweep requires the exchange of columns $v_0 \Leftrightarrow v_2$ and $v_1 \Leftrightarrow v_3$ to create $ST_1(\mathbf{v})$ from $S_1(\mathbf{v})$.

$$ST_1(\mathbf{v}) = \begin{array}{cccc} v_0 & v_1 & v_2 & v_3 \\ \hline - & - & 1 & - \\ - & - & 0 & 0 \end{array} \tag{3.137}$$

The final set $S(\mathbf{v}) \leftarrow S_0(\mathbf{v}) \cap ST_1(\mathbf{v})$ (3.130) is built by an intersection as specified in line 7 of Algorithm 2.

$$S(\mathbf{v}) = \begin{array}{cc} v_0 & v_1 \\ \hline - & - \end{array} \cap \begin{array}{cccc} v_0 & v_1 & v_2 & v_3 \\ \hline - & - & 1 & - \\ - & - & 0 & 0 \end{array} = \begin{array}{cccc} v_0 & v_1 & v_2 & v_3 \\ \hline - & - & 1 & - \\ - & - & 0 & 0 \end{array} \tag{3.138}$$

The set $S(\mathbf{v})$ (3.138) is the result of the second execution of line 7 of Algorithm 2. This set describes 12 solution functions of the BDE (3.107) based on the normal form (3.83) by two ternary vectors.

Fig. 3.7 shows as blue Karnaugh-maps the 12 solution functions of the BDE (3.107) and as black Karnaugh-maps the remaining four functions which do not belong to the solution set. The Karnaugh-maps are ordered in rows such that each row contains all functions of a single function class. Karnaugh-maps colored in gray show the seven representative functions of these classes. It can be seen that four classes contain one function that does not belong to the solution set of the BDE (3.107). These four functions are equal to each other within the subspace $b = 1$ and have in this subspace the monotonously rising function $f(a, b = 1) = a$. The BDE (3.107) specifies that such functions do not belong to the solution set. The split of four classes into solution functions and functions that are not a solution confirms that certain functions out of the classes are separated as solution set of the BDE (3.107) by means of Algorithm 2.

class number	representative function	solution functions $f(a,b)$ of the BDE (3.107)				not a solution of the BDE (3.107)
1	f_0 $\begin{array}{c|cc} b & & \\ 0 & 0 & 0 \\ 1 & 0 & 0 \\ \hline & 0 & 1 \, a \end{array}$	f_0 $\begin{array}{c|cc} b & & \\ 0 & 0 & 0 \\ 1 & 0 & 0 \\ \hline & 0 & 1 \, a \end{array}$				
2	f_1 $\begin{array}{c|cc} b & & \\ 0 & 0 & 0 \\ 1 & 0 & 1 \\ \hline & 0 & 1 \, a \end{array}$	f_1 $\begin{array}{c|cc} b & & \\ 0 & 0 & 0 \\ 1 & 1 & 0 \\ \hline & 0 & 1 \, a \end{array}$	f_2 $\begin{array}{c|cc} b & & \\ 0 & 0 & 1 \\ 1 & 0 & 0 \\ \hline & 0 & 1 \, a \end{array}$	f_3 $\begin{array}{c|cc} b & & \\ 0 & 1 & 0 \\ 1 & 0 & 0 \\ \hline & 0 & 1 \, a \end{array}$		f_4 $\begin{array}{c|cc} b & & \\ 0 & 0 & 0 \\ 1 & 0 & 1 \\ \hline & 0 & 1 \, a \end{array}$
3	f_5 $\begin{array}{c|cc} b & & \\ 0 & 0 & 1 \\ 1 & 0 & 1 \\ \hline & 0 & 1 \, a \end{array}$	f_6 $\begin{array}{c|cc} b & & \\ 0 & 1 & 0 \\ 1 & 1 & 0 \\ \hline & 0 & 1 \, a \end{array}$				f_5 $\begin{array}{c|cc} b & & \\ 0 & 0 & 1 \\ 1 & 0 & 1 \\ \hline & 0 & 1 \, a \end{array}$
4	f_7 $\begin{array}{c|cc} b & & \\ 0 & 0 & 0 \\ 1 & 1 & 1 \\ \hline & 0 & 1 \, a \end{array}$	f_7 $\begin{array}{c|cc} b & & \\ 0 & 0 & 0 \\ 1 & 1 & 1 \\ \hline & 0 & 1 \, a \end{array}$	f_8 $\begin{array}{c|cc} b & & \\ 0 & 1 & 1 \\ 1 & 0 & 0 \\ \hline & 0 & 1 \, a \end{array}$			
5	f_9 $\begin{array}{c|cc} b & & \\ 0 & 0 & 1 \\ 1 & 1 & 0 \\ \hline & 0 & 1 \, a \end{array}$	f_{10} $\begin{array}{c|cc} b & & \\ 0 & 0 & 1 \\ 1 & 1 & 0 \\ \hline & 0 & 1 \, a \end{array}$				f_9 $\begin{array}{c|cc} b & & \\ 0 & 1 & 0 \\ 1 & 0 & 1 \\ \hline & 0 & 1 \, a \end{array}$
6	f_{11} $\begin{array}{c|cc} b & & \\ 0 & 0 & 1 \\ 1 & 1 & 1 \\ \hline & 0 & 1 \, a \end{array}$	f_{12} $\begin{array}{c|cc} b & & \\ 0 & 0 & 1 \\ 1 & 1 & 1 \\ \hline & 0 & 1 \, a \end{array}$	f_{13} $\begin{array}{c|cc} b & & \\ 0 & 1 & 0 \\ 1 & 1 & 1 \\ \hline & 0 & 1 \, a \end{array}$	f_{14} $\begin{array}{c|cc} b & & \\ 0 & 1 & 1 \\ 1 & 1 & 0 \\ \hline & 0 & 1 \, a \end{array}$		f_{11} $\begin{array}{c|cc} b & & \\ 0 & 1 & 1 \\ 1 & 0 & 1 \\ \hline & 0 & 1 \, a \end{array}$
7	f_{15} $\begin{array}{c|cc} b & & \\ 0 & 1 & 1 \\ 1 & 1 & 1 \\ \hline & 0 & 1 \, a \end{array}$	f_{15} $\begin{array}{c|cc} b & & \\ 0 & 1 & 1 \\ 1 & 1 & 1 \\ \hline & 0 & 1 \, a \end{array}$				

Figure 3.7: All 16 functions $f(a,b)$ of B^2 divided into 12 solutions of the BDE (3.107) colored in blue and four functions that are not a solution structured by the seven function classes with representative functions colored in gray.

3.5.2   SEPARATION OF FUNCTIONS USING XBOOLE

The main difference between Algorithm 1 for the separation of function classes and Algorithm 2 for the separation of functions consists in the variables of the solution functions which must be taken into account in Algorithm 2. Hence, the XBOOLE problem programs (PRP) to solve both types of BDEs have a similar structure.

In which way must the **x**-variables of the solution function of a BDE of the type (3.106) be considered when a PRP is used?

1. The **x**-variables of the function $f(\mathbf{x})$ occur in the associated Boolean equation which is solved in line 1 of Algorithm 2 together with the variables **u**. The procedure to solve this associated Boolean equation of a BDE (3.106) must not distinguish between these Boolean variables despite the different meaning of both sets of variables.

2. The function d2v() maps the **u**-variables that describe derivatives onto the **v**-variables of function values. The **x**-variables act in this procedure as parameters which are taken without changes into the result set $S(\mathbf{v}, \mathbf{x})$.

3. Within each sweep of the loop from lines 3 to 8 of Algorithm 2 one variable x_i controls the separation steps. Both the negative cofactor $S_0(\mathbf{v}, \mathbf{x})$ and the positive cofactor $S_1(\mathbf{v}, \mathbf{x})$ of $S(\mathbf{v}, \mathbf{x})$ with regard to x_i are calculated in lines 4 and 5 of Algorithm 2. The max-operations remove the variables x_i from the cofactors.

4. The function epv() must be executed in line 6 of Algorithm 2 only for the positive cofactor $S_1(\mathbf{v}, \mathbf{x})$ of $S(\mathbf{v}, \mathbf{x})$ with regard to x_i.

5. The result of the intersection in line 7 of Algorithm 2 does not depend on the variable x_i, where i is increased in each sweep of the loop. Hence, all **x**-variables are removed from $S(\mathbf{v}, \mathbf{x})$ within the loop of Algorithm 2 so that the result is $S(\mathbf{v})$.

In order to facilitate a direct comparison we solve the BDE (3.107) based on Algorithm 2 by means of a detailed XBOOLE-PRP. Fig. 3.8 shows this PRP.

Recall that the numbers in front of the PRP lines are added in the listing for reference purposes and do not belong to the PRP itself. In line 1 a Boolean space of 32 variables with the number 1 is defined. The used variables are attached to this Boolean space in the wanted order in lines 2 to 5. The associated Boolean equation of the BDE (3.120) is solved in lines 6 and 7 of the listing in Fig. 3.8. These two lines of the PRP realize the task of line 1 in Algorithm 2. The result of this step is the set of local solutions which is stored as an XBOOLE object 1. The content of this object is equal to $SLS(\mathbf{u}, a, b)$ shown in (3.131).

The mapping $S(\mathbf{v}, a, b) \leftarrow \text{d2v}(SLS(\mathbf{u}, a, b))$ of line 2 in Algorithm 2 is realized by lines 8 to 17 of the PRP in Fig. 3.8. The command in line 8 of the PRP solves the set of Boolean equations given in lines 9 to 12 which comply with the system of Boolean equations (3.71) and (3.72) for $n = 2$. The solution set of this system of Boolean equations is the XBOOLE object $XBO[2]$ (3.86)

```
 1   space  32  1                    19   a = 0.
 2   avar  1                         20   isc  5  6  7
 3   u0  u1  u2  u3                   21   maxk  7  6  8
 4   v0  v1  v2  v3                   22   cpl  6  9
 5   a  b.                           23   isc  5  9  10
 6   sbe  1  1                        24   maxk  10  9  11
 7   a&u0&u1+/b=/b.                  25   _cco  11  <v0  v2>  <v1  v3>  12
 8   sbe  1  2                        26   isc  8  12  13
 9   v0=u0 ,                          27   sbe  1  14.
10   v1=u0#u1 ,                       28   b = 0.
11   v2=u0#u2 ,                       29   isc  13  14  15
12   v3=u0#u3.                        30   maxk  15  14  16
13   isc  1  2  3                     31   cpl  14  17
14   vtin  1  4                       32   isc  13  17  18
15   u0  u1  u2  u3.                  33   maxk  18  17  19
16   maxk  3  4  5                    34   _cco  19  <v0  v1>  <v2  v3>  20
17   obb  5  5                        35   isc  16  20  21
18   sbe  1  6.
```

Figure 3.8: Listing of the PRP to solve the BDE (3.107).

which contains the mapping from all 16 Boolean vectors (u_0, u_1, u_2, u_3) to the associated vectors (v_0, v_1, v_2, v_3). The set of transformation vectors depends only on the number of variables of the function in the BDE but not on the BDE itself so that the same XBOOLE object $XBO[2]$ (3.86) is generated as in Subsec. 3.4.4.

The result $XBO[3]$ of the intersection in line 13 consists of 44 vectors $S(\mathbf{u}, \mathbf{v}, a, b)$ which are associated with the 60 local solutions $SLS(\mathbf{u}, a, b)$ of (3.131).

The k-fold maximum in line 16 eliminates all columns (u_0, u_1, u_2, u_3) from $XBO[3]$ and reduces in this way the linear dependency between these variables. The required variables are prepared as variable tuple in lines 14 and 15 of the PRP listing in Fig. 3.8. The 60 potential solution vectors $S(\mathbf{v}, a, b)$ are compressed from 44 to 4 ternary vectors of $XBO[5]$ by the XBOOLE operation obb in line 17.

The XBOOLE object $XBO[5]$ (3.139) contains the same 60 potential solution vectors as $S(\mathbf{v}, a, b)$ (3.132). Columns of dashes for the variables v_2 and v_3 indicate that $S(\mathbf{v}, a, b)$ does not depend on these variables at this point of the solution process.

$$XBO[5] = S(\mathbf{v}, a, b) = \begin{array}{cccccc} v_0 & v_1 & v_2 & v_3 & a & b \\ \hline - & 1 & - & - & 1 & 1 \\ 0 & 0 & - & - & 1 & 1 \\ - & - & - & - & - & 0 \\ - & - & - & - & 0 & 1 \\ \hline \end{array} \qquad (3.139)$$

It remains the separation of the actual solution functions from the set of potential solution functions (3.139). The solution functions of the BDE (3.107) depend on the variables a and b. Hence, the loop of lines 3 to 8 of Algorithm 2 must be executed two times for $i = 1$ and $i = 2$.

In the first sweep of the loop in Algorithm 2 the separation is calculated with regard to the variable a. The negative cofactor $S_0(\mathbf{v}, a = 0, b)$ is calculated by the commands from 18 to 21 as XBOOLE object $XBO[8]$ (3.140), and the positive cofactor $S_1(\mathbf{v}, a = 1, b)$ is created by the commands of lines 22 to 24 as XBOOLE object $XBO[11]$ (3.141).

$$XBO[8] = S_0(\mathbf{v}, b) = \begin{array}{ccccc} v_0 & v_1 & v_2 & v_3 & b \\ \hline - & - & - & - & 0 \\ - & - & - & - & 1 \\ \hline \end{array} \qquad (3.140)$$

$$XBO[11] = S_1(\mathbf{v}, b) = \begin{array}{ccccc} v_0 & v_1 & v_2 & v_3 & b \\ \hline - & 1 & - & - & 1 \\ 0 & 0 & - & - & 1 \\ - & - & - & - & 0 \\ \hline \end{array} \qquad (3.141)$$

The exchange of the columns $v_0 \Leftrightarrow v_1$ and $v_2 \Leftrightarrow v_3$ is only necessary for the positive cofactor. The command in line 25 realizes this exchange of columns from $XBO[11] = S_1(\mathbf{v}, b)$ to $XBO[12] = ST_1(\mathbf{v}, b)$ (3.142).

$$XBO[12] = ST_1(\mathbf{v}, b) = \begin{array}{ccccc} v_0 & v_1 & v_2 & v_3 & b \\ \hline 1 & - & - & - & 1 \\ 0 & 0 & - & - & 1 \\ - & - & - & - & 0 \\ \hline \end{array} \qquad (3.142)$$

The intersection of $XBO[8]$ (3.140) and $XBO[12]$ (3.142) in line 26 of the PRP calculates the XBOOLE object $XBO[13]$ (3.143) which contains the same 26 vectors $(\mathbf{v}, b)$ as $S(\mathbf{v}, b)$ (3.135).

$$XBO[13] = S(\mathbf{v}, b) = \begin{array}{ccccc} v_0 & v_1 & v_2 & v_3 & \\ \hline - & - & - & - & 0 \\ 1 & - & - & - & 1 \\ 0 & 0 & - & - & 1 \\ \hline \end{array} \qquad (3.143)$$

The second sweep of the loop in Algorithm 1 requires the separation with regard to the variable b.

Likewise the negative cofactor $S_0(\mathbf{v}, b = 0)$ is calculated by the commands in lines 27 to 30 as XBOOLE object $XBO[16]$ (3.144), and the positive cofactor $S_1(\mathbf{v}, b = 1)$ is created by the commands in lines 31 to 33 as XBOOLE object $XBO[19]$ (3.145).

$$XBO[16] = S_0(\mathbf{v}) = \begin{array}{cccc} v_0 & v_1 & v_2 & v_3 \\ \hline - & - & - & - \\ \hline \end{array}$$

(3.144)

$$XBO[19] = S_1(\mathbf{v}) = \begin{array}{cccc} v_0 & v_1 & v_2 & v_3 \\ \hline 1 & - & - & - \\ 0 & 0 & - & - \\ \hline \end{array}$$

(3.145)

The exchange of the columns $v_0 \Leftrightarrow v_2$ and $v_1 \Leftrightarrow v_3$ must be executed for the positive cofactor. The command in line 34 realizes this exchange of columns from $XBO[19] = S_1(\mathbf{v})$ to $XBO[20] = ST_1(\mathbf{v})$ (3.146).

$$XBO[20] = ST_1(\mathbf{v}) = \begin{array}{cccc} v_0 & v_1 & v_2 & v_3 \\ \hline - & - & 1 & - \\ - & - & 0 & 0 \\ \hline \end{array}$$

(3.146)

The intersection of $XBO[16]$ (3.144) and $XBO[20]$ (3.146) in line 35 of the PRP calculates the XBOOLE object $XBO[21]$ (3.147) which contains the same two ternary vectors $\mathbf{v}$ as $S(\mathbf{v})$ (3.138). These two ternary vectors express 12 actual solution functions based on the normal form (3.83). The blue Karnaugh-maps in Fig. 3.7 show these solution functions in detail.

$$XBO[21] = S(\mathbf{v}) = \begin{array}{cccc} v_0 & v_1 & v_2 & v_3 \\ \hline - & - & 1 & - \\ - & - & 0 & 0 \\ \hline \end{array}$$

(3.147)

Algorithm 2, like Algorithm 1, uses 2^n variables u_j, $j = 0, \ldots, 2^n - 1$, to describe the function $f(x_1, \ldots, x_n)$ of the BDE and their derivatives and additionally 2^n variables v_j to store the function values of the solution functions. These two exponential sets of variables are required at different steps of the solution procedure. Hence, the restriction to the set of variables v_j, $j = 0, \ldots, 2^n - 1$, which carries information of the **u**-variables at the beginning of the solution process and changes the meaning to the **v**-variables by (3.100) significantly saves resources required in the solution process. This simplified approach of Algorithm 1 can also be used for Algorithm 2 as shown in the following example.

A Boolean function is symmetric with regard to the pair of variables (x_i, x_j) if the exchange of the values assigned to these variables does not change the value of the function for fixed assignments of values to the remaining variables of the function. A symmetric Boolean function satisfies this property for each pair of variables. The BDE

$$\bigvee_{i=2}^{n} \left[(x_1 \oplus x_i) \wedge \frac{\partial f(\mathbf{x})}{\partial (x_1, x_i)} \right] = 0$$

(3.148)

describes the set of symmetric Boolean functions of n variables x_i. In order to save space for the presentation we show in this subsection how the BDE

$$\left[(x_1 \oplus x_2) \wedge \frac{\partial f(x_1, x_2, x_3)}{\partial (x_1, x_2)} \right] \vee \left[(x_1 \oplus x_3) \wedge \frac{\partial f(x_1, x_2, x_3)}{\partial (x_1, x_3)} \right] = 0 \qquad (3.149)$$

for symmetric Boolean functions of three variables x_i can be solved by an XBOOLE problem program. The Boolean equation associated to the BDE (3.149) is:

$$[(x_1 \oplus x_2) \wedge v_3] \vee [(x_1 \oplus x_3) \wedge v_5] = 0 . \qquad (3.150)$$

As mentioned above, the PRP of Fig. 3.9 uses the set of **v**-variables not only to represent the function values of the solution functions but also at the beginning to characterize the function $f(x_1, x_2, x_3)$ and their derivatives. After the definition of the Boolean space in line 1 and the assignment of the variables v_j, $j = 0, \ldots, 7$, and x_i, $i = 1, \ldots, 3$ in lines 2 to 4 the associated Boolean equation (3.150) of the BDE (3.149) is solved in lines 5 and 6. The set of local solutions $SLS(\mathbf{v}, \mathbf{x})$ is stored as XBOOLE object $XBO[1]$ (3.151).

$$XBO[1] = SLS(v_3, v_5, x_1, x_2, x_3) = $$

v_3	v_5	x_1	x_2	x_3
0	–	0	1	0
0	–	1	0	1
–	0	0	0	1
–	–	1	1	1
0	0	0	1	1
–	0	1	1	0
–	–	0	0	0
0	0	1	0	0

$$(3.151)$$

The set of local solutions (3.151) depicts the requirements specified by the BDE (3.149). In all four cases where the value of x_1 is different from the value of x_2 the value of the vectorial derivative of $f(x_1, x_2, x_3)$ with regard to x_1 and x_2 represented by the variables v_3 is equal to zero. As a second requirement the vectorial derivative of $f(x_1, x_2, x_3)$ with regard to x_1 and x_3 represented by the variables v_5 is equal to zero for the four cases that the variables x_1 and x_3 have different values. The required property of symmetry cannot be violated in the remaining two case, where all three variables of the function are equal to each other which is indicated by the dashes for both v_3 and v_5.

The required transformation of line 2 in Algorithm 2 is realized by the change of the meaning of the **v**-variables described by (3.100) in lines 7 to 16 of the PRP shown in Fig. 3.9. The **x**-variables remain unchanged in this procedure with the result of potential solution functions $S(\mathbf{v}, x_1, x_2, x_3)$

```
 1  space  32  1                       30  obb  18  18
 2  avar  1                            31  sbe  1  19.
 3  v0  v1  v2  v3  v4  v5  v6  v7      32  x2 = 0.
 4  x1  x2  x3.                         33  isc  18  19  20
 5  sbe  1  1                          34  maxk  20  19  21
 6  (x1#x2)&v3 + (x1#x3)&v5 = 0.        35  cpl  19  22
 7  sbe  1  2                          36  isc  18  22  23
 8  v0 = 0.                            37  maxk  23  22  24
 9  isc  1  2  3                       38  vtin  1  25
10  cpl  2  4                          39  v0  v1  v4  v5.
11  isc  1  4  5                       40  vtin  1  26
12  vtin  1  6                         41  v2  v3  v6  v7.
13  v1  v2  v3  v4  v5  v6  v7.         42  cco  24  25  26  27
14  cel  5  6  7  /01  /10             43  isc  21  27  28
15  uni  3  7  8                       44  obb  28  28
16  obb  8  8                          45  sbe  1  29.
17  sbe  1  9.                         46  x3 = 0.
18  x1 = 0.                            47  isc  28  29  30
19  isc  8  9  10                      48  maxk  30  29  31
20  maxk  10  9  11                    49  cpl  29  32
21  cpl  9  12                         50  isc  28  32  33
22  isc  8  12  13                     51  maxk  33  32  34
23  maxk  13  12  14                   52  vtin  1  35
24  vtin  1  15                        53  v0  v1  v2  v3.
25  v0  v2  v4  v6.                     54  vtin  1  36
26  vtin  1  16                        55  v4  v5  v6  v7.
27  v1  v3  v5  v7.                     56  cco  34  35  36  37
28  cco  14  15  16  17                57  isc  31  37  38
29  isc  11  17  18                     58  obb  38  38
```

Figure 3.9: Listing of the PRP to solve the BDE (3.149).

which is stored as the XBOOLE object $XBO[8]$ (3.152).

$$XBO[8] = S(v_0, v_3, v_5, x_1, x_2, x_3) = \begin{array}{cccccc} v_0 & v_3 & v_5 & x_1 & x_2 & x_3 \\ \hline 0 & 0 & - & 0 & 1 & 0 \\ 0 & 0 & - & 1 & 0 & 1 \\ 0 & - & 0 & 0 & 0 & 1 \\ 0 & - & - & 1 & 1 & 1 \\ 0 & 0 & 0 & 0 & 1 & 1 \\ 0 & - & 0 & 1 & 1 & 0 \\ 0 & - & - & 0 & 0 & 0 \\ 0 & 0 & 0 & 1 & 0 & 0 \\ 1 & 1 & - & 0 & 1 & 0 \\ 1 & 1 & - & 1 & 0 & 1 \\ 1 & - & 1 & 0 & 0 & 1 \\ 1 & - & - & 1 & 1 & 1 \\ 1 & 1 & 1 & 0 & 1 & 1 \\ 1 & - & 1 & 1 & 1 & 0 \\ 1 & - & - & 0 & 0 & 0 \\ 1 & 1 & 1 & 1 & 0 & 0 \\ \hline \end{array} \qquad (3.152)$$

The function $f(x_1, x_2, x_3)$ of the BDE (3.149) depends on three variables x_i. Hence, the four operations within the loop of Algorithm 2 must be executed three times. The first sweep is realized by the XBOOLE operations in lines 17 to 30 of the PRP shown in Fig. 3.9. The solution of the simple equation $x_1 = 0$ is used by the intersection in line 19 and the maximum operation in line 20 in order to calculate the negative cofactor of $S(\mathbf{v}, x_1, x_2, x_3)$ with regard to x_1. The positive cofactor is built by the complement in line 21, the intersection in line 22, and the maximum operation in line 23. The exchange of columns in the positive cofactor is realized by the XBOOLE operation cco in line 28 of the PRP controlled by the variable tuples which are prepared in lines 24 to 27 of the PRP shown in Fig. 3.9. The intersection in line 29 calculates the final result of the first sweep of the loop. The orthogonal block-building in line 30 reduces the number of ternary vectors of this intermediate result $XBO[18]$ (3.153).

$$XBO[18] = S(\mathbf{v}, x_2, x_3) =$$

v_0	v_1	v_2	v_3	v_4	v_5	x_2	x_3
0	0	–	0	0	–	1	0
0	1	–	0	1	–	1	0
0	0	0	–	–	0	0	1
0	1	1	–	–	0	0	1
0	–	–	0	–	0	1	1
1	0	–	1	0	–	1	0
1	1	–	1	1	–	1	0
1	0	0	–	–	1	0	1
1	1	1	–	–	1	0	1
1	–	–	1	–	1	1	1
–	0	0	–	0	–	0	0
–	1	1	–	1	–	0	0

$$(3.153)$$

The second sweep of the loop is realized by lines 31 to 44 of the PRP. In this sweep the separation is calculated with regard to x_2 and the columns to be changed in the positive cofactor which are specified in the first four rows of the column $i = 2$ of Tab. 3.9. The result of this sweep is stored in the XBOOLE object $XBO[28]$ (3.154) by eight ternary vectors.

$$XBO[28] = S(\mathbf{v}, x_3) =$$

v_0	v_1	v_2	v_3	v_4	v_5	v_6	v_7	x_3
0	0	0	–	–	0	–	0	1
0	1	1	–	–	0	–	1	1
1	0	0	–	–	1	–	0	1
1	1	1	–	–	1	–	1	1
–	0	0	0	0	–	0	–	0
–	0	0	1	0	–	1	–	0
–	1	1	0	1	–	0	–	0
–	1	1	1	1	–	1	–	0

$$(3.154)$$

The actual solution set of the BDE (3.149) is calculated in the third sweep of the loop of Algorithm 2. This sweep is realized by lines 45 to 58 of the PRP shown in Fig. 3.9. In this sweep the separation is calculated with regard to x_3 and the columns to be changed in the positive cofactor which are specified in the first four rows of the column $i = 3$ of Tab. 3.9. The actual set of 16 solution functions $S(\mathbf{v})$ of the BDE (3.149) is stored in the XBOOLE object $XBO[38]$ (3.155) by four ternary vectors.

$$XBO[38] = S(v_0, v_1, v_2, v_3, v_4, v_5, v_6, v_7) =$$

v_0	v_1	v_2	v_3	v_4	v_5	v_6	v_7
–	0	0	0	0	0	0	–
–	0	0	1	0	1	1	–
–	1	1	0	1	0	0	–
–	1	1	1	1	1	1	–

$$(3.155)$$

The set of solution functions $S(\mathbf{v})$ (3.155) can be mapped into the set of actual solution functions $f(x_1, x_2, x_3)$ using the normal form (3.74). The vectors of function values of these functions show their properties quite well. The dashes in the columns v_0 and v_7 of (3.155) mean that arbitrary function values can be chosen for $f(x_1 = 0, x_2 = 0, x_3 = 0)$ and $f(x_1 = 1, x_2 = 1, x_3 = 1)$. The identical values in the columns v_1, v_2, and v_4 mean that function values for $f(x_1 = 1, x_2 = 0, x_3 = 0)$, $f(x_1 = 0, x_2 = 1, x_3 = 0)$, and $f(x_1 = 0, x_2 = 0, x_3 = 1)$ must be identical for symmetric functions of three variables. Similarly, the identical values in the columns v_3, v_5, and v_6 mean that function values for $f(x_1 = 1, x_2 = 1, x_3 = 0)$, $f(x_1 = 1, x_2 = 0, x_3 = 1)$, and $f(x_1 = 0, x_2 = 1, x_3 = 1)$ must be identical for symmetric functions of three variables, too.

3.6 BOOLEAN DIFFERENTIAL EQUATIONS OF ALL DERIVATIVE OPERATIONS

3.6.1 EXAMPLES

Boolean differential equations of the types (3.22) and (3.106) are restricted to derivatives with regard to all directions of change. However, the Boolean Differential Calculus contains many other derivative operations which contribute to the expressiveness of this calculus and the wide field of applications. The following examples of simple BDEs motivate the extension of Algorithm 1 and Algorithm 2 for BDEs in which all types of derivative operations exist.

Boolean functions $f(x_i, \mathbf{x}_1)$ are monotonously rising with regard to x_i if

$$f(x_i = 0, \mathbf{x}_1) \leq f(x_i = 1, \mathbf{x}_1) . \tag{3.156}$$

The solution of the BDE:

$$\overline{x}_i \wedge f(x_i, \mathbf{x}_1) \wedge \frac{\partial f(x_i, \mathbf{x}_1)}{\partial x_i} = 0 \tag{3.157}$$

is the set of monotonously rising functions $f(x_i, \mathbf{x}_1)$ with regard to x_i.

Boolean functions $f(\mathbf{x})$ can have this property for more than one variable. The solution of the BDE

$$f(x_i, x_j, \mathbf{x}_1) \wedge \left(\overline{x}_i \wedge \frac{\partial f(x_i, x_j, \mathbf{x}_1)}{\partial x_i} \vee \overline{x}_j \wedge \frac{\partial f(x_i, x_j, \mathbf{x}_1)}{\partial x_j} \right) = 0 \tag{3.158}$$

is the set of monotonously rising functions $f(x_i, x_j, \mathbf{x}_1)$ with regard to x_i and x_j.

The solution of the BDE (3.158) contains Boolean functions which are in a subspace $\mathbf{x}_1 = \mathbf{c}$ equal to zero:

$$f(x_i, x_j, \mathbf{x}_1) \,|_{\mathbf{x}_1 = \mathbf{c}} = 0 , \tag{3.159}$$

or equal to one:

$$f(x_i, x_j, \mathbf{x}_1) \,|_{\mathbf{x}_1 = \mathbf{c}} = 1 . \tag{3.160}$$

Boolean functions $f(x_i, x_j, \mathbf{x}_1)$ which satisfy (3.158) but not the equation (3.159) for any $\mathbf{x}_1 = \mathbf{c}$ are specified by the BDE:

$$f(x_i, x_j, \mathbf{x}_1) \wedge \left(\overline{x}_i \wedge \frac{\partial f(x_i, x_j, \mathbf{x}_1)}{\partial x_i} \vee \overline{x}_j \wedge \frac{\partial f(x_i, x_j, \mathbf{x}_1)}{\partial x_j} \right) \vee \overline{\max^2_{(x_i, x_j)} f(x_i, x_j, \mathbf{x}_1)} = 0 . \tag{3.161}$$

Vice versa, Boolean functions $f(x_i, x_j, \mathbf{x_1})$ which satisfy (3.158), but not the equation (3.160) for any $\mathbf{x_1} = \mathbf{c}$ are specified by the BDE

$$f(x_i, x_j, \mathbf{x_1}) \wedge \left(\overline{x}_i \wedge \frac{\partial f(x_i, x_j, \mathbf{x_1})}{\partial x_i} \vee \overline{x}_j \wedge \frac{\partial f(x_i, x_j, \mathbf{x_1})}{\partial x_j} \right) \vee \min^2_{(x_i, x_j)} f(x_i, x_j, \mathbf{x_1}) = 0 \; .$$

(3.162)

More generally, the set of monotonously rising functions $f(x_i, x_j, \mathbf{x_1})$ with regard to x_i and x_j which are not constant in the subspace $\mathbf{x_1} = \mathbf{c}$ can be found. These Boolean functions are the solution of the BDE:

$$f(x_i, x_j, \mathbf{x_1}) \wedge \left(\overline{x}_i \wedge \frac{\partial f(x_i, x_j, \mathbf{x_1})}{\partial x_i} \vee \overline{x}_j \wedge \frac{\partial f(x_i, x_j, \mathbf{x_1})}{\partial x_j} \right) \vee \Delta_{(x_i, x_j)} f(x_i, x_j, \mathbf{x_1}) = 0 \; .$$

(3.163)

Boolean differential equations of the explored type can be adapted to the desired requirements. Instead of monotonously rising functions as introduced in (3.156) and specified by the BDE (3.157), monotonously falling functions with regard to x_i

$$f(x_i = 0, \mathbf{x_1}) \geq f(x_i = 1, \mathbf{x_1}) \; .$$

(3.164)

are solutions of the BDE:

$$x_i \wedge f(x_i, \mathbf{x_1}) \wedge \frac{\partial f(x_i, \mathbf{x_1})}{\partial x_i} = 0 \; .$$

(3.165)

Certain required properties can be combined for disjoint subspaces. Boolean functions $f(x_i, x_j, x_k, \mathbf{x_1})$ which satisfy the three properties:

1. monotonously falling with regard to x_i for the subspace $x_j = 0, x_k = 1$,

2. monotonously rising with regard to x_i for the subspace $x_j = x_k$, and

3. constant with regard to x_i for the subspace $x_j = 1, x_k = 0$

are solutions of the BDE:

$$f(x_i, x_j, x_k, \mathbf{x_1}) \wedge \frac{\partial f(x_i, x_j, x_k, \mathbf{x_1})}{\partial x_i} \left(x_i \overline{x}_j x_k \vee \overline{x}_i (x_j \odot x_k) \vee x_j \overline{x}_k \right) = 0 \; .$$

(3.166)

Both the analysis and the synthesis of circuits provide many applications of Boolean differential equations. The behavior of a combinatorial circuit can be described by the system equation:

$$F(\mathbf{x}, \mathbf{y}) = 1 \; .$$

(3.167)

The solution of the Boolean equation (3.167) is the set of binary vectors which specify the valid behavior of the combinatorial circuit. The combinatorial circuit responds for each input pattern

$\mathbf{x} = \mathbf{c}_i$ with a unique output pattern $\mathbf{y} = \mathbf{c}_o$ due to the explicit system of Boolean equations of the associated circuit.

$$y_1 = f_1(\mathbf{x})$$
$$\vdots$$
$$y_m = f_m(\mathbf{x}) \tag{3.168}$$

A first analysis step in a design procedure is the verification whether the behavior of (3.167) is realizable as logic circuit. The circuit is realizable if the equation (3.167) is resolvable with regard to the output variables $\mathbf{y}$. The BDE

$$\max_{\mathbf{y}}^{m} F(\mathbf{x}, \mathbf{y}) = 1 \tag{3.169}$$

describes the set of functions $F(\mathbf{x}, \mathbf{y})$ which are resolvable with regard to $\mathbf{y}$ and therefore realizable as combinatorial circuit (3.168). If a system function satisfies the BDE 3.169, it can be uniquely resolved for a selected output y_i. The BDE

$$\frac{\partial}{\partial y_i} \left(\max_{\mathbf{y} \setminus y_i}^{m-1} F(\mathbf{x}, \mathbf{y}) \right) = 1 \tag{3.170}$$

describes the set of functions $F(\mathbf{x}, \mathbf{y})$ which are uniquely resolvable and therefore realizable with regard to y_i.

Asynchronous sequential circuits are used to process very fast sequences of signals. Glitches on the connection wires can cause errors of required behavior. A glitch is an actual occurrence of a spurious signal of a circuit. A hazard is the possibility that a glitch can occur Steinbach and Posthoff [2010].

There are different types of hazards. All of them are caused by the simultaneous change of two or more signals. We explore as an example both the functional static 0-hazard and the functional static 1-hazard caused by the simultaneous change of the variables x_i and x_j of a Boolean function $f(x_i, x_j, \mathbf{x}_1)$.

All functional static 0-hazards are characterized by the following properties:

- $f(x_i, x_j, \mathbf{x}_1) = 0$ for a certain assignment $x_i, x_j, \mathbf{x}_1 = c_i, c_j, \mathbf{c}_1$,

- $f(x_i, x_j, \mathbf{x}_1) = 0$ for a modified assignment $x_i, x_j, \mathbf{x}_1 = \overline{c}_i, \overline{c}_j, \mathbf{c}_1$, and

- different function values exist in the respective subspace such that the function can switch for a short period in time to the opposite function value 1.

Hence, the set of functions which is free of functional static 0-hazard is the solution of the following BDE:

$$\overline{\max_{(x_i, x_j)} f(x_i, x_j, \mathbf{x}_1)} \wedge \max_{(x_i, x_j)}^{2} f(x_i, x_j, \mathbf{x}_1) = 0 . \tag{3.171}$$

For functional static 1-hazards the possible sequence of function values is $1 \to 0 \to 1$ instead of the expected simple switch $1 \to 1$ of the function $f(x_i, x_j, \mathbf{x_1})$ in the case of the simultaneous change of the values of the variables x_i and x_j. Hence, the BDE

$$\min_{(x_i, x_j)} f(x_i, x_j, \mathbf{x_1}) \wedge \overline{\min^2_{(x_i, x_j)} f(x_i, x_j, \mathbf{x_1})} = 0 \qquad (3.172)$$

has as a solution the set of all $f(x_i, x_j, \mathbf{x_1})$ which have no functional static 1-hazards with regard to the simultaneous change of the values of the variables x_i and x_j.

These two types of functional static hazards must not be explored separately. All Boolean functions $f(x_i, x_j, \mathbf{x_1})$ which have neither a functional static 0-hazard nor a functional static 1-hazard with regard to the simultaneous change of the values of the variables x_i and x_j belong to the solution set of the BDE:

$$\overline{\frac{\partial f(x_i, x_j, \mathbf{x_1})}{\partial(x_i, x_j)}} \wedge \Delta_{(x_i, x_j)} f(x_i, x_j, \mathbf{x_1}) = 0 . \qquad (3.173)$$

All Boolean functions $f(x_i, x_j, \mathbf{x_1})$ which have both one functional static 0-hazard and one functional static 1-hazard with regard to the simultaneous change of the values of the variables x_i and x_j belong to the solution set of the BDE:

$$\overline{\frac{\partial f(x_i, x_j, \mathbf{x_1})}{\partial(x_i, x_j)}} \wedge \Delta_{(x_i, x_j)} f(x_i, x_j, \mathbf{x_1}) = 1 . \qquad (3.174)$$

Boolean differential equations cannot only describe the sets of Boolean functions with functional static hazards but also functional dynamic hazards or even structural hazards. More information to BDEs of these topics are given in Steinbach and Posthoff [2010]. The knowledge about hazards is required to avoid glitches in circuit structures. The aim of BDEs in the field of hazards consists in the decrease of the number of erroneous functions. BDEs can also be used for the design of logic circuits with desirable properties like short delay, low power consumption, or small chip area. The bi-decomposition of Boolean functions is the main method for such applications of BDEs.

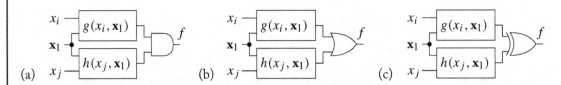

Figure 3.10: Circuit structures of bi-decompositions of the function $f(x_i, x_j, \mathbf{x_1})$ with regard to the variables x_i and x_j: (a) AND-bi-decomposition, (b) OR-bi-decomposition, (c) EXOR-bi-decomposition.

Figure 3.10 shows the circuit structures of the three basic types of bi-decompositions. The benefit of these structures is that the functions $g(x_i, \mathbf{x_1})$ and $h(x_j, \mathbf{x_1})$ depend on one variable less than the output function $f(x_i, x_j, \mathbf{x}_1)$. It is a property of the function $f(x_i, x_j, \mathbf{x}_1)$ whether one type of these bi-decompositions exists.

The set of all $f(x_i, x_j, \mathbf{x}_1)$ which belong to the solution of the BDE:

$$\overline{f(x_i, x_j, \mathbf{x}_1)} \wedge \max_{x_i} f(x_i, x_j, \mathbf{x}_1) \wedge \max_{x_j} f(x_i, x_j, \mathbf{x}_1) = 0 \tag{3.175}$$

are AND-bi-decomposable with regard to the variables x_i and x_j. The duality between the nonlinear operations AND and OR leads to a similar BDE (3.176) for the OR-bi-decomposition.

$$f(x_i, x_j, \mathbf{x}_1) \wedge \max_{x_i} \overline{f(x_i, x_j, \mathbf{x}_1)} \wedge \max_{x_j} \overline{f(x_i, x_j, \mathbf{x}_1)} = 0 \tag{3.176}$$

The EXOR-operation is a linear operation. Hence, the BDE (3.177) that describes the set of Boolean functions $f(x_i, x_j, \mathbf{x}_1)$ which are EXOR-bi-decomposable with regard to the variables x_i and x_j differ significantly from the BDEs (3.175) and (3.176).

$$\frac{\partial^2 f(x_i, x_j, \mathbf{x}_1)}{\partial x_i \, \partial x_j} = 0 \tag{3.177}$$

The BDEs (3.175), (3.176), and (3.177) describe the sets of Boolean functions which are bi-decomposable with regard to the single variables x_i and x_j. The benefit of the bi-decomposition increases when instead of the single variable x_i the subset of variables $\mathbf{x}_a$ and instead of the single variable x_j the subset of variables $\mathbf{x}_b$ of the Boolean function $f(\mathbf{x}_a, \mathbf{x}_b, \mathbf{x}_c)$ can be used because the decomposition function $g(\mathbf{x}_a, \mathbf{x}_c)$ does not depend on the variables $\mathbf{x}_b$ and the decomposition function $h(\mathbf{x}_b, \mathbf{x}_c)$ does not depend on the variables $\mathbf{x}_a$. Hence, both decomposition functions become simpler. The sets of Boolean functions $f(\mathbf{x}_a, \mathbf{x}_b, \mathbf{x}_c)$ which have these more general bi-decompositions with regard to the sets of variables $\mathbf{x}_a = (x_1, \ldots, x_m)$ and $\mathbf{x}_b = (x_{m+1}, \ldots, x_{m+k})$ are described

- for the AND-bi-decomposition by the BDE:

$$\overline{f(\mathbf{x}_a, \mathbf{x}_b, \mathbf{x}_c)} \wedge \max_{\mathbf{x}_a}{}^m f(\mathbf{x}_a, \mathbf{x}_b, \mathbf{x}_c) \wedge \max_{\mathbf{x}_b}{}^m f(\mathbf{x}_a, \mathbf{x}_b, \mathbf{x}_c) = 0 \,, \tag{3.178}$$

- for the OR-bi-decomposition by the BDE:

$$f(\mathbf{x}_a, \mathbf{x}_b, \mathbf{x}_c) \wedge \max_{\mathbf{x}_a}{}^m \overline{f(\mathbf{x}_a, \mathbf{x}_b, \mathbf{x}_c)} \wedge \max_{\mathbf{x}_b}{}^m \overline{f(\mathbf{x}_a, \mathbf{x}_b, \mathbf{x}_c)} = 0 \,, \tag{3.179}$$

- and for the EXOR-bi-decomposition by the BDE:

$$\bigvee_{i=1}^{m} \left(\Delta_{\mathbf{x}_b} \frac{\partial f(x_1, \ldots, x_m, \mathbf{x}_b, \mathbf{x}_c)}{\partial x_i} \right) = 0 \,. \tag{3.180}$$

The introduced bi-decompositions can be extended to a complete design method of logic circuits. The additionally required weak bi-decompositions can also be described by BDEs Posthoff and Steinbach [2004]. A further source of improvement of this design method is that the decomposition is executed not only for a single function $f(\mathbf{x}_a, \mathbf{x}_b, \mathbf{x}_c)$, but for all Boolean functions of a lattice. BDEs that describe this extended case are given in Posthoff and Steinbach [2004], too.

As a last example we return to the BDE (3.93) which describes the most complex bent functions of four variables. In this BDE

$$
\begin{aligned}
&\left(\frac{\partial f(\mathbf{x})}{\partial x_1} \oplus \frac{\partial f(\mathbf{x})}{\partial x_2} \oplus \frac{\partial f(\mathbf{x})}{\partial (x_1, x_2)} \right) \wedge \left(\frac{\partial f(\mathbf{x})}{\partial x_1} \oplus \frac{\partial f(\mathbf{x})}{\partial x_3} \oplus \frac{\partial f(\mathbf{x})}{\partial (x_1, x_3)} \right) \wedge \\
&\left(\frac{\partial f(\mathbf{x})}{\partial x_1} \oplus \frac{\partial f(\mathbf{x})}{\partial x_4} \oplus \frac{\partial f(\mathbf{x})}{\partial (x_1, x_4)} \right) \wedge \left(\frac{\partial f(\mathbf{x})}{\partial x_2} \oplus \frac{\partial f(\mathbf{x})}{\partial x_3} \oplus \frac{\partial f(\mathbf{x})}{\partial (x_2, x_3)} \right) \wedge \\
&\left(\frac{\partial f(\mathbf{x})}{\partial x_2} \oplus \frac{\partial f(\mathbf{x})}{\partial x_4} \oplus \frac{\partial f(\mathbf{x})}{\partial (x_2, x_4)} \right) \wedge \left(\frac{\partial f(\mathbf{x})}{\partial x_3} \oplus \frac{\partial f(\mathbf{x})}{\partial x_4} \oplus \frac{\partial f(\mathbf{x})}{\partial (x_3, x_4)} \right) = 1
\end{aligned}
$$

12 simple derivatives and 6 vectorial derivatives appear. Due to the large number of derivatives it is difficult to understand why just this BDE describes the most complex bent functions of four variables. Another BDE for the same set of most complex bent functions of four variables is:

$$
\frac{\partial^2 f(\mathbf{x})}{\partial x_1 \partial x_2} \wedge \frac{\partial^2 f(\mathbf{x})}{\partial x_1 \partial x_3} \wedge \frac{\partial^2 f(\mathbf{x})}{\partial x_1 \partial x_4} \wedge \frac{\partial^2 f(\mathbf{x})}{\partial x_2 \partial x_3} \wedge \frac{\partial^2 f(\mathbf{x})}{\partial x_2 \partial x_4} \wedge \frac{\partial^2 f(\mathbf{x})}{\partial x_3 \partial x_4} = 1 . \tag{3.181}
$$

It is easier to understand the meaning of (3.181). Bent functions are the functions with the largest distance to all linear Boolean functions of the explored number of variables. That means the bent functions are strongly nonlinear functions. The 2-fold derivative with regard to the variables x_i and x_j describes all Boolean functions which are nonlinear with regard to these variables x_i and x_j. The solutions of the BDE (3.181) are functions which are nonlinear with regard to all pairs of two variables of $\mathbf{x} = (x_1, x_2, x_3, x_4)$.

This example gives a hint to a solution procedure for BDEs of all derivative operations. All derivative operations of the Boolean Differential Calculus can be transformed into expressions in which only simple and vectorial derivatives occur. The comparison of the BDEs shows as example the identity:

$$
\frac{\partial^2 f(\mathbf{x})}{\partial x_1 \partial x_2} = \frac{\partial f(\mathbf{x})}{\partial x_1} \oplus \frac{\partial f(\mathbf{x})}{\partial x_2} \oplus \frac{\partial f(\mathbf{x})}{\partial (x_1, x_2)} . \tag{3.182}
$$

In the following subsections we extend the known methods of Sec. 3.4 and Sec. 3.5 such that each derivative operation can occur in the BDE to be solved. There are different possibilities for mappings between derivative operations. An inductive approach of the required extension of the BDE is presented in the following subsections.

3.6.2 EXTENSION TO ALL SIMPLE DERIVATIVE OPERATIONS

Simple derivatives with regard to all x_i of the function $f(x_i, \mathbf{x}_1)$ and this function itself are usable elements in the algorithm for separation of function classes or arbitrary function sets. Hence, the simple minimum $\min_{x_i} f(x_i, \mathbf{x}_1)$ and the simple maximum $\max_{x_i} f(x_i, \mathbf{x}_1)$ must be mapped to these available elements.

Starting with the definition of the minimum we describe in detail equivalent transformation steps to get an expression of the function $f(x_i, \mathbf{x}_1)$ and their simple derivative with regard to x_i which is equal to the simple minimum $\min_{x_i} f(x_i, \mathbf{x}_1)$.

$$\min_{x_i} f(x_i, \mathbf{x}_1) = f(x_i, \mathbf{x}_1) \wedge f(\overline{x}_i, \mathbf{x}_1) \tag{3.183}$$

$$= f(x_i, \mathbf{x}_1) \, f(x_i, \mathbf{x}_1) \, f(\overline{x}_i, \mathbf{x}_1) \vee f(x_i, \mathbf{x}_1) \, \overline{f(x_i, \mathbf{x}_1)} \, \overline{f(\overline{x}_i, \mathbf{x}_1)} \tag{3.184}$$

$$= f(x_i, \mathbf{x}_1) \wedge \left(f(x_i, \mathbf{x}_1) \, f(\overline{x}_i, \mathbf{x}_1) \vee \overline{f(x_i, \mathbf{x}_1)} \, \overline{f(\overline{x}_i, \mathbf{x}_1)} \right) \tag{3.185}$$

$$= f(x_i, \mathbf{x}_1) \wedge \left(f(x_i, \mathbf{x}_1) \odot f(\overline{x}_i, \mathbf{x}_1) \right) \tag{3.186}$$

$$= f(x_i, \mathbf{x}_1) \wedge \overline{\left(f(x_i, \mathbf{x}_1) \oplus f(\overline{x}_i, \mathbf{x}_1) \right)} \tag{3.187}$$

$$\min_{x_i} f(x_i, \mathbf{x}_1) = f(x_i, \mathbf{x}_1) \wedge \overline{\frac{\partial f(x_i, \mathbf{x}_1)}{\partial x_i}} \tag{3.188}$$

The conjunction of (3.183) is extended by $f(x_i, \mathbf{x}_1)$, and a second conjunction which is equal to 0 due to the law $a \wedge \overline{a} = 0$. The application of the distributive law changes (3.184) into (3.185). The expression in parentheses of (3.185) is equal to the equivalence of $f(x_i, \mathbf{x}_1)$ and $f(\overline{x}_i, \mathbf{x}_1)$ and this equivalence is expressed in (3.187) by the complement of antivalence of these functions. Applying the definition of the derivative to the expression in parentheses of (3.187), the wanted identity (3.188) has been found.

Similar transformation steps facilitate the mapping of the simple maximum of $f(x_i, \mathbf{x}_1)$ with regard to x_i into an expression of this function and their simple derivative with regard to the same variable. A shorter way uses the relation (2.37) between the vectorial maximum and the vectorial minimum applied for the function $f(x_i, \mathbf{x}_1)$ and the single variable x_i as shown in (3.189).

$$\max_{x_i} f(x_i, \mathbf{x}_1) = \overline{\min_{x_i} \overline{f(x_i, \mathbf{x}_1)}} \tag{3.189}$$

$$= \overline{\overline{f(x_i, \mathbf{x}_1)} \wedge \overline{\frac{\partial \overline{f(x_i, \mathbf{x}_1)}}{\partial x_i}}} \tag{3.190}$$

$$= f(x_i, \mathbf{x}_1) \vee \frac{\partial \overline{f(x_i, \mathbf{x}_1)}}{\partial x_i} \tag{3.191}$$

$$\max_{x_i} f(x_i, \mathbf{x}_1) = f(x_i, \mathbf{x}_1) \vee \frac{\partial f(x_i, \mathbf{x}_1)}{\partial x_i} \tag{3.192}$$

The relation (3.188) is substituted on the right-hand side of (3.189). The expression (3.190) is simplified using de Morgan's Law. Using for x_i the property that a function f and the complement $\overline{f}$ have the same derivative, the wanted identity (3.192) has been found.

3.6.3   EXTENSION TO ALL VECTORIAL DERIVATIVE OPERATIONS

All vectorial derivatives with regard to $\mathbf{x}_0$ of the function $f(\mathbf{x}_0, \mathbf{x}_1)$ and this function itself are usable elements in the algorithm for the separation of function classes or for the separation arbitrary function sets. Hence, the vectorial minimum $\min_{\mathbf{x}_0} f(\mathbf{x}_0, \mathbf{x}_1)$ and the vectorial maximum $\max_{\mathbf{x}_0} f(\mathbf{x}_0, \mathbf{x}_1)$ must be mapped to these available elements. The function values of each vectorial derivative operation are calculated based on two function values of the given function as in the case of simple derivative operations. The difference is that all variables of the set $\mathbf{x}_0$ change their values simultaneously. Therefore similar transformations can be used for the mapping of the vectorial minimum and the vectorial maximum to the function $f(\mathbf{x}_0, \mathbf{x}_1)$ and the vectorial derivative with regard to the same variables.

We transform first the vectorial maximum of $f(\mathbf{x}_0, \mathbf{x}_1)$ with regard to $\mathbf{x}_0$ in order to demonstrate alternative transformation steps in comparison to the simple derivative operations.

$$\max_{\mathbf{x}_0} f(\mathbf{x}_0, \mathbf{x}_1) = f(\mathbf{x}_0, \mathbf{x}_1) \vee f(\overline{\mathbf{x}}_0, \mathbf{x}_1) \tag{3.193}$$

$$= f(\mathbf{x}_0, \mathbf{x}_1) \vee \overline{f(\mathbf{x}_0, \mathbf{x}_1)}\, f(\overline{\mathbf{x}}_0, \mathbf{x}_1) \tag{3.194}$$

$$= f(\mathbf{x}_0, \mathbf{x}_1) \vee f(\mathbf{x}_0, \mathbf{x}_1)\, \overline{f(\overline{\mathbf{x}}_0, \mathbf{x}_1)} \vee \overline{f(\mathbf{x}_0, \mathbf{x}_1)}\, f(\overline{\mathbf{x}}_0, \mathbf{x}_1) \tag{3.195}$$

$$= f(\mathbf{x}_0, \mathbf{x}_1) \vee (f(\mathbf{x}_0, \mathbf{x}_1) \oplus f(\overline{\mathbf{x}}_0, \mathbf{x}_1)) \tag{3.196}$$

$$\max_{\mathbf{x}_0} f(\mathbf{x}_0, \mathbf{x}_1) = f(\mathbf{x}_0, \mathbf{x}_1) \vee \frac{\partial f(\mathbf{x}_0, \mathbf{x}_1)}{\partial \mathbf{x}_0} \tag{3.197}$$

Starting with the definition of the vectorial maximum the second term of (3.193) is extended to an orthogonal expression in (3.194). The second term in (3.194) is included based on the absorption law $a \vee a\,b = a$ which is applied in the reverse direction. The last two terms of (3.195) are combined to an antivalence of $f(\mathbf{x}_0, \mathbf{x}_1)$ and $f(\overline{\mathbf{x}}_0, \mathbf{x}_1)$. Applying the definition of the vectorial derivative to the expression in parentheses of (3.196) the wanted identity (3.197) has been found.

Similarly to the case of the mapping of simple derivative operations, we use the relation between the vectorial minimum and the vectorial maximum in order to find the mapping rule for the vectorial minimum of $f(\mathbf{x}_0, \mathbf{x}_1)$ with regard to $\mathbf{x}_0$ to the used function and their vectorial derivative in the same direction.

$$\min_{\mathbf{x}_0} f(\mathbf{x}_0, \mathbf{x}_1) = \overline{\max_{\mathbf{x}_0} \overline{f(\mathbf{x}_0, \mathbf{x}_1)}} \tag{3.198}$$

$$= \overline{\overline{f(\mathbf{x}_0, \mathbf{x}_1)} \vee \frac{\partial \overline{f(\mathbf{x}_0, \mathbf{x}_1)}}{\partial \mathbf{x}_0}} \tag{3.199}$$

$$= \overline{\overline{f(\mathbf{x}_0, \mathbf{x}_1)} \vee \frac{\partial f(\mathbf{x}_0, \mathbf{x}_1)}{\partial \mathbf{x}_0}} \tag{3.200}$$

$$\min_{\mathbf{x}_0} f(\mathbf{x}_0, \mathbf{x}_1) = f(\mathbf{x}_0, \mathbf{x}_1) \wedge \overline{\frac{\partial f(\mathbf{x}_0, \mathbf{x}_1)}{\partial \mathbf{x}_0}} \tag{3.201}$$

The relation (3.197) is substituted on the right-hand side of (3.198). By removing the inner negation in the vectorial derivative in (3.199) the expression (3.200) is simplified using de Morgan's Law and results in the wanted identity (3.201).

3.6.4   EXTENSION TO ALL m-FOLD DERIVATIVE OPERATIONS

The BDEs (3.22) and (3.106) contain only simple and vectorial derivatives. Hence, mapping rules are necessary for m-fold derivative operations. Due to the known transformation rules from Subsec. 3.6.2 and Subsec. 3.6.3 all simple and vectorial derivative operations can be used as basis for these transformations.

Except the Δ-operation, all m-fold derivative operations are defined iteratively. Therefore, we explore the mapping rules in detail for the variables x_i and x_j and generalize them later on for subsets of variables.

$$\frac{\partial^2 f(x_i, x_j, \mathbf{x_1})}{\partial x_i \, \partial x_j} = \frac{\partial}{\partial x_j}\left(\frac{\partial f(x_i, x_j, \mathbf{x_1})}{\partial x_i}\right) \tag{3.202}$$

$$= \frac{\partial}{\partial x_j}\left(f(x_i, x_j, \mathbf{x_1}) \oplus f(\overline{x}_i, x_j, \mathbf{x_1})\right) \tag{3.203}$$

$$= f(x_i, x_j, \mathbf{x_1}) \oplus f(\overline{x}_i, x_j, \mathbf{x_1}) \oplus f(x_i, \overline{x}_j, \mathbf{x_1}) \oplus f(\overline{x}_i, \overline{x}_j, \mathbf{x_1}) \tag{3.204}$$

$$= f(x_i, x_j, \mathbf{x_1}) \oplus f(x_i, x_j, \mathbf{x_1}) \oplus f(x_i, x_j, \mathbf{x_1}) \oplus$$
$$f(\overline{x}_i, x_j, \mathbf{x_1}) \oplus f(x_i, \overline{x}_j, \mathbf{x_1}) \oplus f(\overline{x}_i, \overline{x}_j, \mathbf{x_1}) \tag{3.205}$$

$$\frac{\partial^2 f(x_i, x_j, \mathbf{x_1})}{\partial x_i \, \partial x_j} = \frac{\partial f(x_i, x_j, \mathbf{x_1})}{\partial x_i} \oplus \frac{\partial f(x_i, x_j, \mathbf{x_1})}{\partial x_j} \oplus \frac{\partial f(x_i, x_j, \mathbf{x_1})}{\partial (x_i, x_j)} \tag{3.206}$$

First we search a mapping rule for the 2-fold derivative of $f(x_i, x_j, \mathbf{x_1})$ with regard to x_i and x_j. Using the definition of the single derivative we get the identity (3.202). Simple equivalent transformations facilitate the mapping of the 2-fold derivative into an expression of simple and vectorial derivatives. Using the definition for the simple derivative with regard to x_i in (3.202) we get (3.203). The equivalent ESPO (3.204) is built applying the definition for the simple derivative with regard to x_j in (3.203). Using the rule $a = b \oplus b \oplus a$ the function $f(x_i, x_j, \mathbf{x_1})$ is included twice into (3.204). The pairs of functions which are ordered on top of each other in the two rows of (3.205) satisfy from the left to the right the definition of the simple derivative of $f(x_i, x_j, \mathbf{x_1})$ with regard to x_i, the definition of the simple derivative of $f(x_i, x_j, \mathbf{x_1})$ with regard to x_j, and the definition of the vectorial derivative of $f(x_i, x_j, \mathbf{x_1})$ with regard to (x_i, x_j) as shown in (3.206).

The 2-fold minimum of $f(x_i, x_j, \mathbf{x_1})$ with regard to x_i and x_j can be mapped in a similar way to the available simple and vectorial minima. The definition of the m-fold minimum is applied in (3.207) to $f(x_i, x_j, \mathbf{x_1})$ with regard to the two variables x_i and x_j. The inner simple minimum is executed in (3.208) with regard to x_i and the outer minimum is built in (3.209) with regard to x_j. Using the rule $a = a \wedge a \wedge a$ the function $f(x_i, x_j, \mathbf{x_1})$ is included twice into (3.209). The pairs of functions which are ordered on top of each other in the two rows of (3.210) satisfy from the left to the right the definition of the simple minimum of $f(x_i, x_j, \mathbf{x_1})$ with regard to x_i, the definition of the simple minimum of $f(x_i, x_j, \mathbf{x_1})$ with regard to x_j, and the definition of the vectorial minimum of $f(x_i, x_j, \mathbf{x_1})$ with regard to (x_i, x_j) as shown in (3.211).

$$\min_{(x_i,x_j)}{}^2 f(x_i, x_j, \mathbf{x_1}) = \min_{x_j}\left(\min_{x_i} f(x_i, x_j, \mathbf{x_1})\right) \tag{3.207}$$

$$= \min_{x_j}\left(f(x_i, x_j, \mathbf{x_1}) \wedge f(\overline{x}_i, x_j, \mathbf{x_1})\right) \tag{3.208}$$

$$= f(x_i, x_j, \mathbf{x_1}) \wedge f(\overline{x}_i, x_j, \mathbf{x_1}) \wedge f(x_i, \overline{x}_j, \mathbf{x_1}) \wedge f(\overline{x}_i, \overline{x}_j, \mathbf{x_1}) \tag{3.209}$$

$$= f(x_i, x_j, \mathbf{x_1}) \wedge f(x_i, x_j, \mathbf{x_1}) \wedge f(x_i, x_j, \mathbf{x_1}) \wedge$$
$$f(\overline{x}_i, x_j, \mathbf{x_1}) \wedge f(x_i, \overline{x}_j, \mathbf{x_1}) \wedge f(\overline{x}_i, \overline{x}_j, \mathbf{x_1}) \tag{3.210}$$

$$\min_{(x_i,x_j)}{}^2 f(x_i, x_j, \mathbf{x_1}) = \min_{x_i} f(x_i, x_j, \mathbf{x_1}) \wedge \min_{x_j} f(x_i, x_j, \mathbf{x_1}) \wedge \min_{(x_i,x_j)} f(x_i, x_j, \mathbf{x_1}) \tag{3.211}$$

Due to the duality between the minima and maxima, the 2-fold maximum of $f(x_i, x_j, \mathbf{x_1})$ with regard to x_i and x_j can be mapped in nearly the same way to the available simple and vectorial maxima.

$$\max_{(x_i,x_j)}{}^2 f(x_i, x_j, \mathbf{x_1}) = \max_{x_j}\left(\max_{x_i} f(x_i, x_j, \mathbf{x_1})\right) \tag{3.212}$$

$$= \max_{x_j}\left(f(x_i, x_j, \mathbf{x_1}) \vee f(\overline{x}_i, x_j, \mathbf{x_1})\right) \tag{3.213}$$

$$= f(x_i, x_j, \mathbf{x_1}) \vee f(\overline{x}_i, x_j, \mathbf{x_1}) \vee f(x_i, \overline{x}_j, \mathbf{x_1}) \vee f(\overline{x}_i, \overline{x}_j, \mathbf{x_1}) \tag{3.214}$$

$$= f(x_i, x_j, \mathbf{x_1}) \vee f(x_i, x_j, \mathbf{x_1}) \vee f(x_i, x_j, \mathbf{x_1}) \vee$$
$$f(\overline{x}_i, x_j, \mathbf{x_1}) \vee f(x_i, \overline{x}_j, \mathbf{x_1}) \vee f(\overline{x}_i, \overline{x}_j, \mathbf{x_1}) \tag{3.215}$$

$$\max_{(x_i,x_j)}{}^2 f(x_i, x_j, \mathbf{x_1}) = \max_{x_i} f(x_i, x_j, \mathbf{x_1}) \vee \max_{x_j} f(x_i, x_j, \mathbf{x_1}) \vee \max_{(x_i,x_j)} f(x_i, x_j, \mathbf{x_1}) \tag{3.216}$$

The definition of the m-fold maximum is applied in (3.212) to $f(x_i, x_j, \mathbf{x_1})$ with regard to the two variables x_i and x_j. The inner simple maximum is executed in (3.213) with regard to x_i and the outer maximum is built in (3.214) with regard to x_j. Using the rule $a = a \vee a \vee a$ the function $f(x_i, x_j, \mathbf{x_1})$ is included twice into (3.214). The pairs of functions which are ordered on top of each other in the two rows of (3.215) satisfy from the left to the right the definiton of the simple maximum of $f(x_i, x_j, \mathbf{x_1})$ with regard to x_i, the definition of the simple maximum of $f(x_i, x_j, \mathbf{x_1})$ with regard to x_j, and the definition of the vectorial maximum of $f(x_i, x_j, \mathbf{x_1})$ with regard to (x_i, x_j) as shown in (3.216).

Based on the sequences of mappings of the 2-fold minimum and the 2-fold maximum to simple and vectorial derivatives, the last missing mapping rule for $\Delta_{(x_i,x_j)} f(x_i, x_j, \mathbf{x_1})$ can be specified.

$$\Delta_{(x_i,x_j)} f(x_i, x_j, \mathbf{x_1}) = \min_{(x_i,x_j)}{}^2 f(x_i, x_j, \mathbf{x_1}) \oplus \max_{(x_i,x_j)}{}^2 f(x_i, x_j, \mathbf{x_1}) \tag{3.217}$$

The mapping rule (3.206), (3.211), and (3.216) will now be extended using the power set $P(\mathbf{x_0})$. The power set $P(\mathbf{x_0})$ is the set of all non-empty subsets of variables of the set of variables $\mathbf{x_0}$.

Iterative extensions of the mapping rules (3.206), (3.211), and (3.216) lead for $\mathbf{x}_0 = (x_1, x_2, \ldots, x_m)$ directly to the following equations:

$$\frac{\partial^m f(\mathbf{x}_0, \mathbf{x}_1)}{\partial x_1 \, \partial x_2 \ldots \partial x_m} = \bigoplus_{\mathbf{y} \in P(\mathbf{x}_0)} \frac{\partial f(\mathbf{x}_0, \mathbf{x}_1)}{\partial \mathbf{y}} , \tag{3.218}$$

$$\min_{\mathbf{x}_0}^m f(\mathbf{x}_0, \mathbf{x}_1) = \bigwedge_{\mathbf{y} \in P(\mathbf{x}_0)} \min_{\mathbf{y}} f(\mathbf{x}_0, \mathbf{x}_1) , \tag{3.219}$$

$$\max_{\mathbf{x}_0}^m f(\mathbf{x}_0, \mathbf{x}_1) = \bigvee_{\mathbf{y} \in P(\mathbf{x}_0)} \max_{\mathbf{y}} f(\mathbf{x}_0, \mathbf{x}_1) . \tag{3.220}$$

The mapping of the m-fold minimum or the m-fold maximum into derivative operations of a BDE of the types (3.22) or (3.106) needs two steps of transformations. The definition of the Δ-operation results in an expression that includes both the m-fold minimum and the m-fold maximum. Hence, the mapping of the Δ-operation into derivative operations of a BDE of the types (3.22) or (3.106) needs even a sequence of three transformation steps.

Direct mapping rules for all m-fold derivative operations of $f(\mathbf{x}_0, \mathbf{x}_1)$ with regard to the variables $\mathbf{x}_0 = (x_1, x_2, \ldots, x_m)$ use the associated power set $P(\mathbf{x}_0)$ and are the results of:

- the direct use of (3.218) for the m-fold derivative,

- the Th. (2.54) for the Δ-operation,

- the substitution of (3.188) and (3.201) into (3.219) for the m-fold minimum, and

- the substitution of (3.192) and (3.197) into (3.220) for the m-fold maximum.

$$\frac{\partial^m f(\mathbf{x}_0, \mathbf{x}_1)}{\partial x_1 \, \partial x_2 \ldots \partial x_m} = \bigoplus_{\mathbf{y} \in P(\mathbf{x}_0)} \frac{\partial f(\mathbf{x}_0, \mathbf{x}_1)}{\partial \mathbf{y}} \tag{3.221}$$

$$\Delta_{\mathbf{x}_0} f(\mathbf{x}_0, \mathbf{x}_1) = \bigvee_{\mathbf{y} \in P(\mathbf{x}_0)} \frac{\partial f(\mathbf{x}_0, \mathbf{x}_1)}{\partial \mathbf{y}} \tag{3.222}$$

$$\min_{\mathbf{x}_0}^m f(\mathbf{x}_0, \mathbf{x}_1) = f(\mathbf{x}_0, \mathbf{x}_1) \wedge \bigwedge_{\mathbf{y} \in P(\mathbf{x}_0)} \overline{\frac{\partial f(\mathbf{x}_0, \mathbf{x}_1)}{\partial \mathbf{y}}} \tag{3.223}$$

$$\max_{\mathbf{x}_0}^m f(\mathbf{x}_0, \mathbf{x}_1) = f(\mathbf{x}_0, \mathbf{x}_1) \vee \bigvee_{\mathbf{y} \in P(\mathbf{x}_0)} \frac{\partial f(\mathbf{x}_0, \mathbf{x}_1)}{\partial \mathbf{y}} \tag{3.224}$$

3.6.5 METHODS OF TRANSFORMATIONS AS PREPROCESSING STEP

Simple and vectorial derivatives occur in BDEs of the types (3.22) or (3.106) in negated or unnegated form. Hence, these derivatives are expressed in the associated Boolean equation by the Boolean variables $\mathbf{u} = (u_0, \ldots, u_{2^n-1})$. All other derivative operations have the same property. Therefore, these additional derivative operations can also be modeled by Boolean variables. The $\mathbf{u}$-variables are required in the algorithms for the separation of function classes or for the separation of arbitrary function sets. In order to distinguish the $\mathbf{u}$-variables from Boolean variables which express the other derivative operations of a BDE we map all derivatives which do not belong to $\nabla f(\mathbf{x})$ (3.67) to Boolean variables $\mathbf{w}$. For the simple case of $n = 2$ variables x_i and x_j the set of variables $\mathbf{u}$ consists of the variables (u_0, u_1, u_2, u_3), and the other derivative operations can be modeled as follows:

$$
\begin{aligned}
w_1 &\Leftarrow \min_{x_i} f(x_i, x_j, \mathbf{x}_1) \\
w_2 &\Leftarrow \min_{x_j} f(x_i, x_j, \mathbf{x}_1) \\
w_3 &\Leftarrow \min_{(x_i x_j)} f(x_i, x_j, \mathbf{x}_1) \\
w_4 &\Leftarrow \max_{x_i} f(x_i, x_j, \mathbf{x}_1) \\
w_5 &\Leftarrow \max_{x_j} f(x_i, x_j, \mathbf{x}_1) \\
w_6 &\Leftarrow \max_{(x_i x_j)} f(x_i, x_j, \mathbf{x}_1) \\
w_7 &\Leftarrow \min_{(x_i x_j)}{}^2 f(x_i, x_j, \mathbf{x}_1) \\
w_8 &\Leftarrow \max_{(x_i x_j)}{}^2 f(x_i, x_j, \mathbf{x}_1) \\
w_9 &\Leftarrow \Delta_{(x_i x_j)} f(x_i, x_j, \mathbf{x}_1) \\
w_{10} &\Leftarrow \frac{\partial^2 f(x_i, x_j, \mathbf{x}_1)}{\partial x_i \, \partial x_j} .
\end{aligned}
\tag{3.225}
$$

Using (3.225), the mapping rules (3.188), (3.192), (3.197), (3.201), (3.206), (3.211), (3.216), and (3.217), all derivatives which do not belong to $\nabla f(\mathbf{x})$ (3.67) can be expressed by the system of Boolean equations (3.226):

$$
\begin{aligned}
w_1 &= u_0 \wedge /u_1 \\
w_2 &= u_0 \wedge /u_2 \\
w_3 &= u_0 \wedge /u_3 \\
w_4 &= u_0 \vee u_1 \\
w_5 &= u_0 \vee u_2 \\
w_6 &= u_0 \vee u_3 \\
w_7 &= w_1 \wedge w_2 \wedge w_3 \\
w_8 &= w_4 \vee w_5 \vee w_6 \\
w_9 &= w_7 \oplus w_8 \\
w_{10} &= u_1 \oplus u_2 \oplus u_3 .
\end{aligned}
\tag{3.226}
$$

$$T(\mathbf{u}, \mathbf{w}) = \begin{array}{cccc|cccccccccc}
u_0 & u_1 & u_2 & u_3 & w_1 & w_2 & w_3 & w_4 & w_5 & w_6 & w_7 & w_8 & w_9 & w_{10} \\
\hline
0 & 1 & 1 & 1 & 0 & 0 & 0 & 1 & 1 & 1 & 0 & 1 & 1 & 1 \\
1 & 1 & 1 & 1 & 0 & 0 & 0 & 1 & 1 & 1 & 0 & 1 & 1 & 1 \\
0 & 0 & 0 & 1 & 0 & 0 & 0 & 0 & 0 & 1 & 0 & 1 & 1 & 1 \\
1 & 0 & 0 & 1 & 1 & 1 & 0 & 1 & 1 & 1 & 0 & 1 & 1 & 1 \\
0 & 0 & 1 & 0 & 0 & 0 & 0 & 0 & 1 & 0 & 0 & 1 & 1 & 1 \\
1 & 0 & 1 & 0 & 1 & 0 & 1 & 1 & 1 & 1 & 0 & 1 & 1 & 1 \\
1 & 1 & 0 & 0 & 0 & 1 & 1 & 1 & 1 & 1 & 0 & 1 & 1 & 1 \\
0 & 1 & 0 & 0 & 0 & 0 & 0 & 1 & 0 & 0 & 0 & 1 & 1 & 1 \\
0 & 0 & 1 & 1 & 0 & 0 & 0 & 0 & 1 & 1 & 0 & 1 & 1 & 0 \\
1 & 0 & 1 & 1 & 1 & 0 & 0 & 1 & 1 & 1 & 0 & 1 & 1 & 0 \\
0 & 1 & 1 & 0 & 0 & 0 & 0 & 1 & 1 & 0 & 0 & 1 & 1 & 0 \\
1 & 1 & 1 & 0 & 0 & 0 & 1 & 1 & 1 & 1 & 0 & 1 & 1 & 0 \\
0 & 1 & 0 & 1 & 0 & 0 & 0 & 1 & 0 & 1 & 0 & 1 & 1 & 0 \\
1 & 1 & 0 & 1 & 0 & 1 & 0 & 1 & 1 & 1 & 0 & 1 & 1 & 0 \\
1 & 0 & 0 & 0 & 1 & 1 & 1 & 1 & 1 & 1 & 1 & 1 & 0 & 0 \\
0 & 0 & 0 & 0 & 0 & 0 & 0 & 0 & 0 & 0 & 0 & 0 & 0 & 0 \\
\end{array} \qquad (3.227)$$

The solution of the system of Boolean equations (3.226) is the set of binary vectors with the characteristic function $T(\mathbf{u}, \mathbf{w})$ (3.227). This set of 16 binary vectors summarizes the relationships between all derivative operations of $f(x_i, x_j, \mathbf{x}_1)$ with regard to x_i and x_j.

The function $T(\mathbf{u}, \mathbf{w})$ (3.227) facilitates the *single step transformation* as preprocessing method to solve a BDE (3.22) in which all derivative operations of $f(x_i, x_j, \mathbf{x}_1)$ with regard to x_i and x_j can occur. The associated Boolean equation of such a BDE depends on the variables $\mathbf{u}$ and $\mathbf{w}$, and has the set of requirements $SR(\mathbf{u}, \mathbf{w})$ as a solution. The set of local solutions $SLS(\mathbf{u})$ of Algorithm 1 can be calculated by:

$$SLS(\mathbf{u}) = \max_{\mathbf{w}}^{m} [SR(\mathbf{u}, \mathbf{w}) \wedge T(\mathbf{u}, \mathbf{w})] \ . \qquad (3.228)$$

The same method can be used to solve BDEs (3.106) which are extended to all derivative operations. The associated Boolean equation depends in this case on the variables $\mathbf{u}, \mathbf{w}$, and $\mathbf{x}$, and has the set of requirements $SR(\mathbf{u}, \mathbf{w}, \mathbf{x})$ as solution. The set of local solutions $SLS(\mathbf{u}, \mathbf{x})$ of Algorithm 2 can be calculated by:

$$SLS(\mathbf{u}, \mathbf{x}) = \max_{\mathbf{w}}^{m} [SR(\mathbf{u}, \mathbf{w}, \mathbf{x}) \wedge T(\mathbf{u}, \mathbf{w})] \ . \qquad (3.229)$$

This preprocessing step can be easily included into an XBOOLE PRP. As an example we solve the BDE (3.230) which describes the set of very simple functions $f(x_i, x_j)$ without any static 0-hazards with regard to x_i and x_j.

$$\overline{\max_{(x_i, x_j)} f(x_i, x_j)} \wedge \max_{(x_i, x_j)}^{2} f(x_i, x_j) = 0 \qquad (3.230)$$

Using the stipulated assignments (3.225) the associated Boolean equation of (3.230) is:

$$\overline{w}_6 \wedge w_8 = 0 \, . \tag{3.231}$$

```
 1   space 32 1                      20   vtin 1 3
 2   avar 1                          21   w1 w2 w3 w4 w5
 3   u0 u1 u2 u3                     22   w6 w7 w8 w9 w10 .
 4   v0 v1 v2 v3                     23   isc 1 2 4
 5   w1 w2 w3 w4 w5                  24   maxk 4 3 5
 6   w6 w7 w8 w9 w10 .               25   obb 5 5
 7   sbe 1 1                         26   sbe 1 6
 8   /w6&w8 = 0.                     27   v0 = u0 ,
 9   sbe 1 2                         28   v1 = u0#u1 ,
10   w1 = u0&/u1 ,                   29   v2 = u0#u2 ,
11   w2 = u0&/u2 ,                   30   v3 = u0#u3 .
12   w3 = u0&/u3 ,                   31   isc 5 6 7
13   w4 = u0+u1 ,                    32   _maxk 7 <u0 u1 u2 u3> 8
14   w5 = u0+u2 ,                    33   _cco 8 <v0 v2> <v1 v3> 9
15   w6 = u0+u3 ,                    34   isc 8 9 10
16   w7 = w1&w2&w3 ,                 35   _cco 10 <v0 v1> <v2 v3> 11
17   w8 = w4+w5+w6 ,                 36   isc 10 11 12
18   w9 = w7#w8 ,                    37   obb 12 12
19   w10 = u1#u2#u3 .
```

Figure 3.11: Listing of the PRP to solve the BDE (3.230).

Figure 3.11 shows the PRP to solve the BDE (3.230) using the *single step transformation* as preprocessing method of the separation of function classes with Algorithm 1. After the definition of the Boolean space in line 1 the used sets of variables $\mathbf{u}$, $\mathbf{v}$, and $\mathbf{w}$ are assigned to this space in lines 2 to 6. The associated Boolean equation (3.231) is solved in lines 7 and 8 and has the very simple solution $XBO[1]$ (3.232).

$$XBO[1] = SR(\mathbf{w}) = \begin{array}{cc} w_6 & w_8 \\ 1 & 1 \\ \hline - & 0 \end{array} \tag{3.232}$$

The system of Boolean equations in lines 10 to 22 is taken from (3.226). The solution $XBO[2]$ of this system of equations is the set of binary vectors as shown in (3.227).

The single step transformation (3.228) is calculated in lines 23 to 25 of the PRP in Figure 3.11. The XBOOLE object $XBO[5]$ is the set of local solutions $SLS(\mathbf{u})$ (3.233) which is required in line 2 of Algorithm 1. The XBOOLE objects $XBO[1]$ (3.232) and $XBO[5]$ (3.233) describe the same set of solution functions utilizing different elements of the Boolean Differential Calculus.

The comparison between the TVLs of $XBO[1]$ (3.232) and $XBO[5]$ (3.233) reveals the benefits of the different facilities of the Boolean Differential Calculus to express the requirements of a certain function set.

$$XBO[5] = SLS(\mathbf{u}) = \begin{array}{cccc} u_0 & u_1 & u_2 & u_3 \\ \hline 1 & 1 & - & 0 \\ 1 & 0 & 1 & - \\ 0 & 0 & 1 & 1 \\ - & 1 & - & 1 \\ - & 0 & 0 & - \\ \hline \end{array} \qquad (3.233)$$

Lines 26 to 37 of the PRP specify the steps from line 2 to line 6 of Algorithm 1 starting with the $\mathbf{u} \to \mathbf{v}$ mapping in lines 26 to 32 of the PRP and followed by two separation steps in lines 33 to 36. The XBOOLE operation obb in line 37 increases the number of ternary vectors to express the set of solution functions $XBO[12] = S(\mathbf{v})$ of the BDE (3.230) based on the normal form (3.74).

$$XBO[12] = S(\mathbf{v}) = \begin{array}{cccc} v_0 & v_1 & v_2 & v_3 \\ \hline 1 & 1 & 0 & - \\ - & 0 & 1 & 1 \\ 0 & 1 & - & 1 \\ 1 & 1 & 1 & 1 \\ 1 & - & 1 & 0 \\ 0 & 0 & 0 & 0 \\ \hline \end{array} \qquad (3.234)$$

The solution set $S(\mathbf{v})$ describes by six ternary vectors 10 functions $f(x_i, x_j)$ which do not contain any functional static 0-hazard with regard to the simultaneous change of x_i and x_j. These 10 functions belong to five different function classes:

$$
\begin{aligned}
1: \quad & f_0(x_i, x_j) = 0, \\
2: \quad & f_1(x_i, x_j) = x_i, & f_2(x_i, x_j) = \overline{x}_i, \\
3: \quad & f_3(x_i, x_j) = x_j, & f_4(x_i, x_j) = \overline{x}_j, \\
4: \quad & f_5(x_i, x_j) = x_i \vee x_j, & f_6(x_i, x_j) = \overline{x}_i \vee x_j, \\
& f_7(x_i, x_j) = x_i \vee \overline{x}_j, & f_8(x_i, x_j) = \overline{x}_i \vee \overline{x}_j, \\
5: \quad & f_9(x_i, x_j) = 1.
\end{aligned} \qquad (3.235)
$$

The drawback of the *single step transformation* is that the number of $\mathbf{u}$-variables and $\mathbf{w}$-variables grows exponentially with the number n of $\mathbf{x}$-variables of the wanted solution functions. Hence, this method of BDE transformation is restricted to BDEs of functions of few variables.

$$|\mathbf{u}| + |\mathbf{w}| = 7 * 2^n - 4 * n - 6 \qquad (3.236)$$

An alternative transformation method is the *multiple step transformation*. Instead of the transformation of all possible derivative operations only the existing associated variables which do not represent a derivative operation of $\nabla f(\mathbf{x})$ are mapped separately to the associated variables of $\nabla f(\mathbf{x})$

within a loop. This transformation method needs dedicated Boolean equations for the direct mapping of a single existing derivative operation. The required mapping rules are defined by (3.188), (3.192), (3.197), (3.201), (3.221), (3.222), (3.223), and (3.224). In order to distinguish both transformation methods, we use in the associated Boolean equation of a BDE indiviual assigned variables z_i as models of derivative operations which do not belong to $\nabla f(\mathbf{x})$. Each transformation step uses the individual transformation function $f_{ti}(z_l, \mathbf{u})$ in the following equation:

$$SR(\mathbf{u}, \mathbf{z} \setminus z_i) = \max_{z_i} [SR(\mathbf{u}, \mathbf{z}) \wedge f_{ti}(z_i, \mathbf{u})] \ . \tag{3.237}$$

We apply the method of *multiple step transformation* in connection with the separation of function classes to the calculation of all bent functions of four variables. In Steinbach and Posthoff [2011] the BDE:

$$\left(\frac{\partial^2 f(\mathbf{x})}{\partial x_1 \partial x_2} \wedge \frac{\partial^2 f(\mathbf{x})}{\partial x_3 \partial x_4} \right) \oplus \left(\frac{\partial^2 f(\mathbf{x})}{\partial x_1 \partial x_3} \wedge \frac{\partial^2 f(\mathbf{x})}{\partial x_2 \partial x_4} \right) \oplus \left(\frac{\partial^2 f(\mathbf{x})}{\partial x_1 \partial x_4} \wedge \frac{\partial^2 f(\mathbf{x})}{\partial x_2 \partial x_3} \right) = 1 \tag{3.238}$$

of this function set was published. The associated Boolean equation is:

$$(z_1 \wedge z_2) \oplus (z_3 \wedge z_4) \oplus (z_5 \wedge z_6) = 1 \ , \tag{3.239}$$

where due to (3.221) and Def. 3.5 the following dedicated mapping equation must be used within the loop of the *multiple step transformation*:

$$z_1 = u_1 \oplus u_2 \oplus u_3 \ , \tag{3.240}$$
$$z_2 = u_4 \oplus u_8 \oplus u_{12} \ , \tag{3.241}$$
$$z_3 = u_1 \oplus u_4 \oplus u_5 \ , \tag{3.242}$$
$$z_4 = u_2 \oplus u_8 \oplus u_{10} \ , \tag{3.243}$$
$$z_5 = u_1 \oplus u_8 \oplus u_9 \ , \tag{3.244}$$
$$z_6 = u_2 \oplus u_4 \oplus u_6 \ . \tag{3.245}$$

The benefits of the multiple step transformation already become visible for this example of a BED of the function $f(x_1, x_2, x_3, x_4)$ of $n = 4$ variables x_i. Instead of $|\mathbf{u}| + |\mathbf{w}| = 90$ only 16 **u**-variables plus six **z**-variables are necessary.

Figure 3.12 shows the PRP to solve the BDE (3.238) using the *multiple step transformation* as a preprocessing method of the separation of function classes with Algorithm 1. The associated Boolean equation (3.239) is solved in lines 8 and 9 of the PRP and consists of 28 solutions stored in 13 ternary vectors of the XBOOLE object $XBO[1]$.

The **v**-variables in the PRP of Figure 3.12 are used in the same way as shown in the PRP of Figure 3.4 to express first the meaning of the **u**-variables and change the meaning to the **v**-variables during the mapping process in lines 34 to 40 based on (3.100). For that reason the variables $z_i, i = 1, \ldots, 6$ are transformed directly to the **v**-variables within six dedicated transformation steps in lines 10 to 33.

```
 1   space 32 1                          37   isc 19 20 22
 2   avar 1                              38   cel 19 19 23 /01 /10
 3   v00 v01 v02 v03                     39   isc 23 21 24
 4   v04 v05 v06 v07                     40   uni 22 24 25
 5   v08 v09 v10 v11                     41   vtin 1 26
 6   v12 v13 v14 v15                     42   v00 v02 v04 v06
 7   z1 z2 z3 z4 z5 z6 .                 43   v08 v10 v12 v14 .
 8   sbe 1 1                             44   vtin 1 27
 9   z1&z2#z3&z4#z5&z6 = 1.              45   v01 v03 v05 v07
10   sbe 1 2                             46   v09 v11 v13 v15 .
11   z1 = v01#v02#v03 .                  47   cco 25 26 27 28
12   isc 1 2 3                           48   isc 25 28 29
13   _maxk 3 < z1 > 4                    49   vtin 1 30
14   sbe 1 5                             50   v00 v01 v04 v05
15   z2 = v04#v08#v12 .                  51   v08 v09 v12 v13 .
16   isc 4 5 6                           52   vtin 1 31
17   _maxk 6 < z2 > 7                    53   v02 v03 v06 v07
18   sbe 1 8                             54   v10 v11 v14 v15 .
19   z3 = v01#v04#v05 .                  55   cco 29 30 31 32
20   isc 7 8 9                           56   isc 29 32 33
21   _maxk 9 < z3 > 10                   57   vtin 1 34
22   sbe 1 11                            58   v00 v01 v02 v03
23   z4 = v02#v08#v10 .                  59   v08 v09 v10 v11 .
24   isc 10 11 12                        60   vtin 1 35
25   _maxk 12 < z4 > 13                  61   v04 v05 v06 v07
26   sbe 1 14                            62   v12 v13 v14 v15 .
27   z5 = v01#v08#v09 .                  63   cco 32 34 35 36
28   isc 13 14 15                        64   isc 33 36 37
29   _maxk 15 < z5 > 16                  65   vtin 1 38
30   sbe 1 17                            66   v00 v01 v02 v03
31   z6 = v02#v04#v06 .                  67   v04 v05 v06 v07 .
32   isc 16 17 18                        68   vtin 1 39
33   _maxk 18 < z6 > 19                  69   v08 v09 v10 v11
34   sbe 1 20                            70   v12 v13 v14 v15 .
35   v00 = 0 .                           71   cco 37 38 39 40
36   cpl 20 21                           72   isc 37 40 41
```

Figure 3.12: Listing of the PRP to solve the BDE (3.238).

The function $f(\mathbf{x})$ of the BDE (3.238) depends on four variables x_i. Hence, the operations of the loop of Algorithm 1 are repeated four times controlled by the $\mathbf{v}$-variables as specified in (3.73). The solution of the BDE (3.238) consists of 896 Boolean functions $f(x_1, x_2, x_3, x_4)$ which are stored as XBOOLE object $XBO[41]$ as a result of the PRP of Figure 3.12 with the interpretation of the normal form (3.74). These 896 solutions are structured in 56 function classes of 16 Boolean functions. Representatives of these function classes are given both as Boolean expressions and depicted as Karnaugh-maps in Steinbach and Posthoff [2011].

3.7   MOST GENERAL BOOLEAN DIFFERENTIAL EQUATIONS

In all BDEs so far explored only one Boolean function $f(\mathbf{x})$ appears for which the set of solutions is wanted. However, several different unknown functions can also appear within a BDE. In the case of two different unknown functions $f(\mathbf{x})$ and $g(\mathbf{x})$ in a BDE it can be verified for each pair of functions $\langle f_i(\mathbf{x}), g_j(\mathbf{x})\rangle$ whether this pair of functions satisfies the BDE. Hence, such a BDE has a set of function pairs as solution. The generalization of this consideration for a BDE of k different unknown functions $f(\mathbf{x})[j]$, $j = 1, \ldots, k$ has a solution set in which each element consists of k Boolean functions $f(\mathbf{x})[j]$.

There are two problems to be solved for a BDE of k different unknown functions $f(\mathbf{x})[j]$, $j = 1, \ldots, k$:

1. How can the set of solution functions be found for each of these unknown functions $f(\mathbf{x})[j]$?

2. Which combinations of these sets of functions solve the given BDE?

The methods to solve a BDE of a single unknown function $f(\mathbf{x})$ by the separation of function classes or by the separation of arbitrary function sets and the optional preprocessing transformation steps can be adapted to a BDE of several unknown functions $f(\mathbf{x})[j]$, $j = 1, \ldots, k$. The derivative operations of each function $f(\mathbf{x})[j]$ are transformed in a preprocessing step to an individual vector $\nabla f(\mathbf{x})[j]$. Consequently, the associated Boolean equation depends on k separate sets of variables $\mathbf{u}[j]$, $j = 1, \ldots, k$ which are mapped to k separate sets of variables $\mathbf{v}[j]$, $j = 1, \ldots, k$. The valid combinations of the solution functions $f(\mathbf{x})[j]$ are preserved from the BDE to the final solution set when the required manipulations are executed completely for all sets of variables $\mathbf{v}[j]$ within each step of the used separation algorithm.

We apply this method to the solution of a BDE of two unknown functions of a practical problem of the circuit design. The decomposition of a Boolean function in simpler subfunctions is an efficient method for the design of multiple-level circuits. It depends on the properties of the function whether such a decomposition exists. The condition for the AND-bi-decomposition of $f(x_i, x_j, \mathbf{x}_1)$ with regard to x_i and x_j is (3.175). The possibility to find an AND-bi-decomposition grows when not only the function $f(x_i, x_j, \mathbf{x}_1)$ but any function out of a lattice of functions can be used. A lattice of functions can be described by the ON-set $q(x_i, x_j, \mathbf{x}_1)$ and the OFF-set

$r(x_i, x_j, \mathbf{x_1})$ such that all functions $f_l(x_i, x_j, \mathbf{x_1})$ of the lattice satisfy:

$$q(x_i, x_j, \mathbf{x_1}) \leq f_l(x_i, x_j, \mathbf{x_1}) \leq \overline{r(x_i, x_j, \mathbf{x_1})} \, . \tag{3.246}$$

Each pair of functions $\langle q(x_i, x_j, \mathbf{x_1}), \ r(x_i, x_j, \mathbf{x_1}) \rangle$ which satisfies the condition:

$$q(x_i, x_j, \mathbf{x_1}) \wedge r(x_i, x_j, \mathbf{x_1}) = 0 \tag{3.247}$$

describes one lattice of functions. The BDE (3.175) can be generalized to

$$r(x_i, x_j, \mathbf{x_1}) \wedge \max_{x_i} q(x_i, x_j, \mathbf{x_1}) \wedge \max_{x_j} q(x_i, x_j, \mathbf{x_1}) = 0 \tag{3.248}$$

which describes the set of all lattices which contain at least one AND-bi-decomposable function $f(x_i, x_j, \mathbf{x_1})$.

It can be seen in (3.248) that the decomposition property must be satisfied for all $\mathbf{x_1} = \mathbf{c}$. Hence, we restrict ourselves to the BDE

$$r(x_i, x_j) \wedge \max_{x_i} q(x_i, x_j) \wedge \max_{x_j} q(x_i, x_j) = 0 \tag{3.249}$$

of two simple unknown functions $q(x_i, x_j)$ and $r(x_i, x_j)$. The BDE (3.249) satisfies implicitly the lattice condition (3.247). We label the two different sets of $\mathbf{u}, \mathbf{v}$, and $\mathbf{z}$ variables by the second letter q or r which indicate the respective function. The associated Boolean function of the BDE (3.249) is:

$$ur_0 \wedge zq_1 \wedge zq_2 = 1 \, , \tag{3.250}$$

with

$$zq_1 = uq_0 \vee uq_1 \, , \tag{3.251}$$
$$zq_2 = uq_0 \vee uq_2 \, . \tag{3.252}$$

The PRP in Figure 3.13 solves the BDE (3.249) in the following sequence of subtasks:

- line 1: definition of the Boolean space,

- lines 2 to 7: assignment of the Boolean variables,

- lines 8 and 9: solution of the associated Boolean equation,

- lines 10 to 13: first steps of transformation: $\mathbf{uq} \leftarrow zq_1$,

- lines 14 to 17: second steps of transformation $\mathbf{uq} \leftarrow zq_2$,

- lines 18 to 24: mappings $\mathbf{vq} \leftarrow \mathbf{uq}$,

- lines 25 to 31: mappings $\mathbf{vr} \leftarrow \mathbf{ur}$,

```
 1   space 32 1                        20   vq1=uq0#uq1 ,
 2   avar 1                            21   vq2=uq0#uq2 ,
 3   uq0 uq1 uq2 uq3                   22   vq3=uq0#uq3 .
 4   ur0 ur1 ur2 ur3                   23   isc 7 8 9
 5   vq0 vq1 vq2 vq3                   24   _maxk 9 <uq0 uq1 uq2 uq3> 10
 6   vr0 vr1 vr2 vr3                   25   sbe 1 11
 7   zq1 zq2 .                         26   vr0=ur0 ,
 8   sbe 1 1                           27   vr1=ur0#ur1 ,
 9   ur0&zq1&zq2 =0.                   28   vr2=ur0#ur2 ,
10   sbe 1 2                           29   vr3=ur0#ur3 .
11   zq1=uq0+uq1 .                     30   isc 10 11 12
12   isc 1 2 3                         31   _maxk 12 <ur0 ur1 ur2 ur3> 13
13   _maxk 3 <zq1> 4                   32   _cco 13 <vq0 vq2> <vq1 vq3> 14
14   sbe 1 5                           33   _cco 14 <vr0 vr2> <vr1 vr3> 15
15   zq2=uq0+uq2 .                     34   isc 13 15 16
16   isc 4 5 6                         35   _cco 16 <vq0 vq1> <vq2 vq3> 17
17   _maxk 6 <zq2> 7                   36   _cco 17 <vr0 vr1> <vr2 vr3> 18
18   sbe 1 8                           37   isc 16 18 19
19   vq0=uq0 ,                         38   obbc 19 20
```

Figure 3.13: Listing of the PRP to solve the BDE (3.249).

- lines 32 to 34: first sweep of the separation loop with regard to the first variable x_i,

- lines 35 to 37: second sweeps of the separation loop with regard to the first variable x_j, and

- line 38: minimize the solution set to less ternary vectors.

The BDE (3.249) has 71 pairs of functions $\langle q(x_i, x_j), r(x_i, x_j)\rangle$ as solution which are stored in 12 ternary vectors. These pairs of functions describe 71 different lattices which include at least one function $f(x_i, x_j)$ that is AND-bi-decomposable with regard to x_i and x_j. In the solution set $S(\mathbf{vq}, \mathbf{vr}) = XBO[20]$, it can be easily verified that all pairs of functions $\langle q(x_i, x_j), r(x_i, x_j)\rangle$ satisfy the lattice condition: each 1 or dash (-) of vq_i meets a 0 of vr_i and vice versa.

$$XBO[20] = S(\mathbf{vq}, \mathbf{vr}) = \begin{array}{cccccccc} vq_0 & vq_1 & vq_2 & vq_3 & vr_0 & vr_1 & vr_2 & vr_3 \\ \hline - & 1 & 1 & - & 0 & 0 & 0 & 0 \\ 1 & 0 & - & 1 & 0 & 0 & 0 & 0 \\ 1 & 1 & 0 & 1 & 0 & 0 & 0 & 0 \\ - & - & 0 & 0 & 0 & 0 & - & - \\ - & 0 & 0 & 0 & 0 & 1 & - & - \\ 0 & - & 0 & 0 & 1 & 0 & - & - \\ 0 & 1 & 0 & 1 & - & 0 & - & 0 \\ 0 & 0 & 0 & 1 & - & - & 1 & 0 \\ 0 & 0 & 0 & 0 & 1 & 1 & - & - \\ 0 & 0 & - & 1 & - & - & 0 & 0 \\ 0 & 0 & 1 & 0 & 1 & - & 0 & - \\ - & 0 & 1 & 0 & 0 & - & 0 & - \end{array}$$

(3.253)

SUMMARY

In contrast to Boolean equations which have sets of Boolean vectors as solutions, the solution of a Boolean differential equation consists of a set of Boolean functions. This is based on the fact that the result of any selected derivative operation can be the same for different given functions. Hence, Boolean differential equations facilitate the work with sets of Boolean functions in a common way.

More interesting than the reverse operation of a selected derivative operation of the Boolean Differential Calculus is the solution of more general Boolean differential equations in which several derivative operations of the Boolean Differential Calculus appear. The key to solve such general Boolean differential equations is the utilization of relationships between Boolean functions and all simple and vectorial derivatives. The solution of each Boolean differential equation in which only simple and vectorial derivatives are connected by Boolean operations is a class of Boolean functions. Additional Boolean variables in such a differential equation facilitate the description of each set of Boolean functions.

Utilizing theorems and relationships of the Boolean Differential Calculus, all derivative operations can be transformed into simple and vectorial derivatives of a function and this function itself. The approaches to solve a Boolean differential equation of one single unknown function can be easily extended to Boolean differential equations in which several unknown functions appear. Hence, there is no theoretical restriction to solve each Boolean differential equation.

XBOOLE is a powerful practical tool that can be utilized in the field of Boolean differential equations, too. XBOOLE does not limit the number of variables; our focus on function sets of few variables is motivated by the restricted space of this book. Several Boolean differential equations are successfully applied in circuit design and cryptography. There is a wide open field for applications in the future.

EXERCISES

3.1 How many solutions function have the BDEs:

$$\frac{\partial f(a, b, c)}{\partial a} = ((\overline{a}\, b \vee b\, c) \oplus b(a \oplus c)) , \tag{3.254}$$

$$\frac{\partial f(a, b, c)}{\partial a} = ((\overline{a}\, b \oplus b\, c) \oplus b(a \oplus c)) . \tag{3.255}$$

Substantiate your solution.

3.2 Specify a BDE that describes all functions $f(a, b)$ which are linear with regard to the variable a. Verify whether the found BDE is solvable and calculate all solution functions using (3.7) and (3.6).

3.3 Verify that only the functions $g(a, b)$ enumerated in the last column of Tab. 3.2 hold the integrability condition (3.15).

3.4 Specify a BDE that describes all self-dual functions $f(a, b)$. Each self-dual function $f(a, b)$ holds the property $f(a, b) = \overline{f(\overline{a}, \overline{b})}$. Verify whether the found BDE is solvable and calculate all solution functions using (3.17) and (3.16).

3.5 It is shown in (3.42) and (3.43) that $f(a, b) = a \odot b$ and $f(a, b) = a \oplus b$ are solution functions of the BDE (3.25). The single counter-example (3.44) verifies only that the function $f(a, b) = a \wedge b$ does not satisfies the BDE (3.25). Verify by the remaining 13 Boolean function $f(a, b)$ that no other function $f(a, b)$ satisfies the BDE (3.25).

3.6 Bent functions are such Boolean functions which have the largest Hamming distance to all linear functions. Bent functions are needed for cryptographic applications. The solution of the Boolean differential equation:

$$\frac{\partial f(a, b)}{\partial a} \oplus \frac{\partial f(a, b)}{\partial b} = \frac{\partial f(a, b)}{\partial (a, b)} \tag{3.256}$$

is the set of all bent functions of two variables. Solve the BDE (3.256) using the detailed method of Subsec. 3.4.1.

3.7 Figure 3.2 shows the construction of the solution function $f(a, \overline{b}) = a \vee \overline{b}$ from a known solution function $f(a, b) = a \vee b$. Which local solutions of the BDEs (3.46), (3.47), (3.48), and (3.49) describe the remaing two functions of the solution class of the BDE (3.50)?

3.8 How often must the loop of Algorithm 1 be executed in order to solve the BDE:

$$\frac{\partial f(x_1, x_2, x_3, x_4, x_5)}{\partial x_1} \oplus \frac{\partial f(x_1, x_2, x_3, x_4, x_5)}{\partial x_3} = \frac{\partial f(x_1, x_2, x_3, x_4, x_5)}{\partial (x_2, x_5)} . \tag{3.257}$$

3.9 Modify the PRP of Figure 3.3 such that similar to the PRP of Figure 3.4 only the variables v_0, v_1, v_2, and v_3 are used to solve the BDE (3.25). Additionally the number of used XBOOLE objects should be reduced by reusing of the XBOOLE objects within the loop for the separation of the solution classes. Take into account the XBOOLE command _cco which is described in the help system of the XBOOLE-Monitor. Execute the created PRP by means of the XBOOLE-Monitor and verify the calculated solution.

3.10 The BDE (3.107) describes the set of Boolean functions which are monotonously rising in the direction of a for $b = 1$. Modify the BDE (3.107) such that the new BDE describes the set of Boolean functions $f(a, b)$ which are both monotonously falling in the direction of a for $b = 1$ and monotonously rising in the direction of a for $b = 0$. Solve this new BDE using Algorithm 2 of Subsec. 3.5.1 and count the number of solution functions $f(a, b)$.

3.11 Extend the PRP of Figure 3.9 such that the BDE (3.148) for symmetric Boolean functions of four variables ($n = 4$) will be solved. How many symmetric Boolean functions $f(x_1, x_2, x_3, x_4)$ exist?

3.12 Which functions $f(x_i, x_j)$ have both a functional static 0-hazard and a functional static 1-hazard? Solve the BDE which describes this function set using the single step transformation in combination with the separation of function classes.

3.13 The BDE:

$$\frac{\partial^2 f(\mathbf{x})}{\partial x_1 \partial x_2} \wedge \frac{\partial^2 f(\mathbf{x})}{\partial x_3 \partial x_4} \wedge \overline{\frac{\partial^2 f(\mathbf{x})}{\partial x_1 \partial x_3}} \wedge \overline{\frac{\partial^2 f(\mathbf{x})}{\partial x_2 \partial x_4}} \wedge \overline{\frac{\partial^2 f(\mathbf{x})}{\partial x_1 \partial x_4}} \wedge \overline{\frac{\partial^2 f(\mathbf{x})}{\partial x_2 \partial x_3}} = 1 \qquad (3.258)$$

of a subset of the simplest bent function of four variables x_i is given in Steinbach and Posthoff [2011]. Modify the PRP of Figure 3.12 such that the BDE (3.258) is solved. How many solution functions does the BDE (3.258) have?

3.14 Calculate the set of all lattices which include at least one OR-bi-decomposable function $f(x_i, x_j)$ by means of a BDE. Which BED describes this set of lattices? A modified PRP of Figure 3.13 can be used to solve the BDE. How many such lattices exist?

CHAPTER 4

Solutions of the Exercises

4.1 SOLUTION OF CHAPTER 1

1.1 Both sides of the equation can be equivalently transformed as follows:

$$\overline{x_1 \odot x_2} = \overline{x}_1 \oplus \overline{x}_2 \, ,$$
$$x_1 \oplus x_2 = x_1 \oplus 1 \oplus x_2 \oplus 1 \, ,$$
$$x_1 \oplus x_2 = x_1 \oplus x_2 \oplus 0 \, ,$$
$$x_1 \oplus x_2 = x_1 \oplus x_2 \, .$$

de Morgan's Law is valid for the operations $\odot$ and $\oplus$.

1.2 Both the formula F_1 and the formula F_2 describe the same function $f = \overline{a} \vee c$.

1.3 The solution of the Boolean equation (1.28) consists of five binary vectors (a, b, c): {000, 010, 011, 101, 111}. The Boolean equation (1.28) is solvable with regard to the variable a and with regard to the variable c, but not with regard to the variable b because there is no solution of the equation for $a = 1$ and $c = 0$. The solution functions are:

$$a = f_{11}(b, c) = c \, ,$$
$$a = f_{12}(b, c) = \overline{b} c \, ,$$
$$c = f_{31}(a, b) = a \, ,$$
$$c = f_{32}(a, b) = a \vee b \, .$$

1.4 Both the command cco and the command _cco exchange any number of needed pairs of columns within one sweep. The columns to exchange are defined by two variable tuples (VT): the first variables in each of these VTs specify the first pair of columns which have to be exchanged, the second variables specify the second pair of columns and so on. Hence, the exchange of the columns $a \leftrightarrow c$ and $b \leftrightarrow d$ requires the variable tuples $< a\,b >$ and $< c\,d >$. The command cco requires that these VTs are stored as XBOOLE objects which can be used by the commands

```
vtin 1 3
a b.
vtin 1 4
c d.
```

The XBOOLE command

```
cco 1 3 4 2
```

creates the XBOOLE object $XBO[2]$ (4.1):

$$XBO[2] = \begin{array}{cccc} a & b & c & d \\ \hline 1 & 1 & 0 & - \\ 1 & 0 & 1 & 1 \\ 0 & - & - & 1 \\ \hline \end{array} \tag{4.1}$$

using $XBO[1]$ (1.29) and controlled by the VT which are stored as $XBO[3]$ and $XBO[4]$. The XBOOLE command

```
_cco 1 <a b> <c d> 2
```

contains the VTs to control the column exchange directly as parameter. The command cco should be preferred when the VTs are already stored in previous calculation steps or the same VTs are needed several times.

1.5 The result of the execution of the PRP are TVLs stored as XBOOLE objects 1, 2, 3, and 4. Theses for TVLs can be visualized as shown in Fig. 1.8.

4.2   SOLUTION OF CHAPTER 2

2.1 $\max_y F(a, b, c, y) = 1$; hence, Eq. (2.76) is solvable with regard to y and a realizable circuit exists.

$\frac{\partial F(a,b,c,y)}{\partial y} = (a \odot b) \vee c \neq 1$; hence, Eq. (2.76) is not uniquely solvable with regard to y so that several realizable circuits exist.

$\varphi(a, b, c) = \max_y F(a, b, c, y) = (a \oplus b) \vee \overline{c}$; hence, $\varphi(a, b, c)$ is equal to 1 for two patterns $(a, b, c) = (010, 100)$ so that $2^2 = 4$ different functions $y = f(a, b, c)$ satisfy the permitted behavior (2.76). The four solution functions are:

$$\begin{aligned} f_1(a, b, c) &= (a \oplus b) \vee c, \\ f_2(a, b, c) &= \overline{a} b \vee c, \\ f_3(a, b, c) &= a \overline{b} \vee c, \\ f_4(a, b, c) &= c. \end{aligned}$$

2.2 The vectorial derivative of the function (2.77) with regard to all variables **x** is equal to 1. Hence, the function (2.77) is a self-dual function.

2.3 The function $f(a, b, c, d)$ (2.77) is constant in subspaces $(a, b) = const$ if the Δ-operation with regard to the variables (c, d) is equal to 0.

$$\Delta_{(c,d)} f(a, b, c, d) = a \odot b. \tag{4.2}$$

Hence, the function (2.77) is constant in the subspace $(a = 0, b = 1)$ and in the subspace $(a = 1, b = 0)$.

2.4 The 2-fold derivative of $f(a, b, c, d)$ (2.77) with regard to a and d is:

$$\frac{\partial^2 f(a, b, c, d)}{\partial a\, \partial d} = b \odot c \neq 0 . \tag{4.3}$$

Hence, the function $f(a, b, c, d)$ (2.77) cannot be EXOR-decomposed into the functions $g(a, b, c)$ and $h(b, c, d)$.

4.3   SOLUTION OF CHAPTER 3

3.1 The BDE (3.254) has no solution function because the function $g(a, b, c) = b\,(a \vee \overline{c})$ of the right-hand side does not hold the integrability condition.

The BDE (3.255) has $2^{2^{3-1}} = 16$ solution functions because the function $f(a, b, c)$ depends on three variables and the function $g(a, b, c) = b$ of the right-hand side holds the integrability condition.

3.2 The searched BDE is

$$\frac{\partial f(a, b)}{\partial a} = 1 . \tag{4.4}$$

The integrability condition (3.5) holds for $g(a, b) = 1$. There are four solution functions:

$$\begin{aligned}
f_1(a, b) &= a , \\
f_2(a, b) &= a \oplus b , \\
f_3(a, b) &= a \oplus \overline{b} , \\
f_4(a, b) &= \overline{a} .
\end{aligned} \tag{4.5}$$

3.3 The calculation of the vectorial derivative of all 16 functions of the two variables a and b with regard to (a, b) verifies the statement. It can be seen in Tab. 3.2 that the result of the vectorial derivative of exactly each of the functions $\{0, a \oplus b, a \odot b, 1\}$ is equal to 0.

3.4 The searched BDE is

$$\frac{\partial f(a, b)}{\partial (a, b)} = 1 . \tag{4.6}$$

The integrability condition (3.15) holds for $g(a, b) = 1$. There are four solution functions:

$$\begin{aligned}
f_1(a, b) &= a , \\
f_2(a, b) &= b , \\
f_3(a, b) &= \overline{a} , \\
f_4(a, b) &= \overline{b} ,
\end{aligned} \tag{4.7}$$

3.5 The BDE (3.25) can be transformed into

$$\frac{\partial f(a,b)}{\partial a} \vee \frac{\partial f(a,b)}{\partial b} = \overline{\frac{\partial f(a,b)}{\partial(a,b)}}$$

$$\left(f(a,b) \oplus f(\overline{a},b)\right) \vee \left(f(a,b) \oplus f(a,\overline{b})\right) = 1 \oplus f(a,b) \oplus f(\overline{a},\overline{b}) \qquad (4.8)$$

using the definitions of the simple and vectorial derivatives. The remaining 13 functions $f(a,b)$ are substituted into (4.8).

$f(a,b) = 0$:

$$(0 \oplus 0) \vee (0 \oplus 0) \stackrel{?}{=} 1 \oplus 0 \oplus 0$$
$$0 \neq 1 \qquad (4.9)$$

$f(a,b) = a\overline{b}$:

$$\left(a\overline{b} \oplus \overline{a}\,\overline{b}\right) \vee \left(a\overline{b} \oplus ab\right) \stackrel{?}{=} 1 \oplus a\overline{b} \oplus \overline{a}b$$
$$\overline{b} \vee a \neq a \odot b \qquad (4.10)$$

$f(a,b) = a$:

$$(a \oplus \overline{a}) \vee (a \oplus a) \stackrel{?}{=} 1 \oplus a \oplus \overline{a}$$
$$1 \vee 0 \neq 0 \qquad (4.11)$$

$f(a,b) = \overline{a}b$:

$$(\overline{a}b \oplus ab) \vee \left(\overline{a}b \oplus \overline{a}\,\overline{b}\right) \stackrel{?}{=} 1 \oplus \overline{a}b \oplus a\overline{b}$$
$$b \vee \overline{a} \neq a \odot b \qquad (4.12)$$

$f(a,b) = b$:

$$(b \oplus b) \vee \left(b \oplus \overline{b}\right) \stackrel{?}{=} 1 \oplus b \oplus \overline{b}$$
$$0 \vee 1 \neq 0 \qquad (4.13)$$

$f(a,b) = a \vee b = a \oplus b \oplus ab$:

$$(a \oplus b \oplus ab \oplus \overline{a} \oplus b \oplus \overline{a}b) \vee$$
$$\left(a \oplus b \oplus ab \oplus a \oplus \overline{b} \oplus a\overline{b}\right) \stackrel{?}{=} 1 \oplus a \oplus b \oplus ab \oplus \overline{a} \oplus \overline{b} \oplus \overline{a}\,\overline{b}$$
$$\overline{b} \vee \overline{a} \neq a \oplus b \qquad (4.14)$$

$f(a,b) = \overline{a}\,\overline{b}$:

$$\left(\overline{a}\,\overline{b} \oplus a\overline{b}\right) \vee \left(\overline{a}\,\overline{b} \oplus \overline{a}b\right) \stackrel{?}{=} 1 \oplus \overline{a}\,\overline{b} \oplus ab$$
$$\overline{b} \vee \overline{a} \neq a \oplus b \qquad (4.15)$$

$f(a, b) = \bar{b}$:

$$\left(\bar{b} \oplus \bar{b}\right) \vee \left(\bar{b} \oplus b\right) \overset{?}{=} 1 \oplus \bar{b} \oplus b$$
$$0 \vee 1 \neq 0 \tag{4.16}$$

$f(a, b) = a \vee \bar{b} = a \oplus \bar{b} \oplus a\bar{b}$:

$$\left(a \oplus \bar{b} \oplus a\bar{b} \oplus \bar{a} \oplus \bar{b} \oplus \bar{a}\bar{b}\right) \vee$$
$$\left(a \oplus \bar{b} \oplus a\bar{b} \oplus a \oplus b \oplus ab\right) \overset{?}{=} 1 \oplus a \oplus \bar{b} \oplus a\bar{b} \oplus \bar{a} \oplus b \oplus \bar{a}b$$
$$b \vee \bar{a} \neq a \odot b \tag{4.17}$$

$f(a, b) = \bar{a}$:

$$\left(\bar{a} \oplus a\right) \vee \left(\bar{a} \oplus \bar{a}\right) \overset{?}{=} 1 \oplus \bar{a} \oplus a$$
$$1 \vee 0 \neq 0 \tag{4.18}$$

$f(a, b) = \bar{a} \vee b = \bar{a} \oplus b \oplus \bar{a}b$:

$$\left(\bar{a} \oplus b \oplus \bar{a}b \oplus a \oplus b \oplus ab\right) \vee$$
$$\left(\bar{a} \oplus b \oplus \bar{a}b \oplus \bar{a} \oplus \bar{b} \oplus \bar{a}\bar{b}\right) \overset{?}{=} 1 \oplus \bar{a} \oplus b \oplus \bar{a}b \oplus a \oplus \bar{b} \oplus a\bar{b}$$
$$\bar{b} \vee a \neq a \odot b \tag{4.19}$$

$f(a, b) = \bar{a} \vee \bar{b} = \bar{a} \oplus \bar{b} \oplus \bar{a}\bar{b}$:

$$\left(\bar{a} \oplus \bar{b} \oplus \bar{a}\bar{b} \oplus a \oplus \bar{b} \oplus a\bar{b}\right) \vee$$
$$\left(\bar{a} \oplus \bar{b} \oplus \bar{a}\bar{b} \oplus \bar{a} \oplus b \oplus \bar{a}b\right) \overset{?}{=} 1 \oplus \bar{a} \oplus \bar{b} \oplus \bar{a}\bar{b} \oplus a \oplus b \oplus ab$$
$$b \vee a \neq a \oplus b \tag{4.20}$$

$f(a, b) = 1$:

$$(1 \oplus 1) \vee (1 \oplus 1) \overset{?}{=} 1 \oplus 1 \oplus 1$$
$$0 \neq 1 \tag{4.21}$$

None of the remaining 13 functions satisfies the BDE (3.25). Hence, the detailed method of Sec. 3.4 found the complete set of solution functions $f_1(a, b) = a \odot b$ and $f_2(a, b) = a \oplus b$ of the BDE (3.25).

3.6 The solution set of the BDE (3.256) contains eight functions:

$$f_1(a, b) = a \wedge b,$$
$$f_2(a, b) = a \wedge \bar{b},$$
$$f_3(a, b) = \bar{a} \wedge b,$$
$$f_4(a, b) = \bar{a} \wedge \bar{b},$$
$$f_5(a, b) = a \vee b,$$
$$f_6(a, b) = a \vee \bar{b},$$
$$f_7(a, b) = \bar{a} \vee b,$$
$$f_8(a, b) = \bar{a} \vee \bar{b}.$$

(4.22)

3.7 The solution function $f(a, b) = \bar{a} \vee b$ of the BDE (3.50) is specified by the local solutions:

(3.46) for $a = 0$, $b = 1$,
(3.47) for $a = 1$, $b = 0$,
(3.48) for $a = 0$, $b = 0$, and
(3.49) for $a = 1$, $b = 1$.

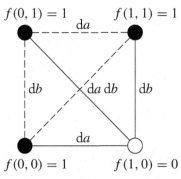

The solution function $f(a, b) = \bar{a} \vee \bar{b}$ of the BDE (3.50) is specified by the local solutions:

(3.46) for $a = 1$, $b = 0$,
(3.47) for $a = 1$, $b = 1$,
(3.48) for $a = 0$, $b = 1$, and
(3.49) for $a = 0$, $b = 0$.

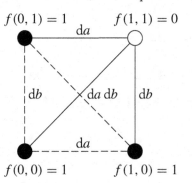

3.8 The loop of Algorithm 1 must be executed five times because the function of the BDE (3.257) depends on five Boolean variables. It doesn't matter that no derivative with regard to x_4 occurs in this BDE.

3.9 The short PRP to solve the BDE (3.25) is:

```
1   space 32 1              9   cpl 2 2
2   avar 1                 10   cel 1 1 4 /01 /10
3   v0 v1 v2 v3.           11   isc 2 4 4
4   sbe 1 1                12   uni 3 4 4
5   v1+v2 =/v3.            13   _cco 4 <v0 v2> <v1 v3> 3
6   sbe 1 2.              14   isc 4 3 4
7   v0 =0.                15   _cco 4 <v0 v1> <v2 v3> 3
8   isc 1 2 3             16   isc 4 3 4
```

and calculates the set of solution functions $S(\mathbf{v})$ as XBO[4]:

$$XBO[4] = S(\mathbf{v}) = \begin{array}{cccc} v_0 & v_1 & v_2 & v_3 \\ \hline 0 & 1 & 1 & 0 \\ 1 & 0 & 0 & 1 \end{array} \tag{4.23}$$

that contains the same vectors of function values as already shown in (3.82) and (3.92).

3.10 Boolean functions $f(a, b)$ which are both monotonously falling in the direction of a for $b = 1$ and monotonously rising in the direction of a for $b = 0$ are solutions of the BDE

$$a \wedge b \wedge f(a,b) \wedge \frac{\partial f(a,b)}{\partial a} \vee \bar{a} \wedge \bar{b} \wedge f(a,b) \wedge \frac{\partial f(a,b)}{\partial a} = 0 \tag{4.24}$$

which can be simplified to

$$(a \odot b) \wedge f(a,b) \wedge \frac{\partial f(a,b)}{\partial a} = 0 . \tag{4.25}$$

There are nine solution functions of the BDEs (4.24) or (4.25) described by $S(\mathbf{v})$ (4.26) and the normal form (3.83).

$$S(\mathbf{v}) = \begin{array}{cccc} v_0 & v_1 & v_2 & v_3 \\ \hline - & 1 & 1 & - \\ - & 1 & 0 & 0 \\ 0 & 0 & 1 & - \\ 0 & 0 & 0 & 0 \end{array} \tag{4.26}$$

3.11 The BDE

$$\left((x_1 \oplus x_2) \wedge \frac{\partial f(x_1, x_2, x_3, x_4)}{\partial(x_1, x_2)}\right) \vee \left((x_1 \oplus x_3) \wedge \frac{\partial f(x_1, x_2, x_3, x_4)}{\partial(x_1, x_3)}\right) \vee \left((x_1 \oplus x_4) \wedge \frac{\partial f(x_1, x_2, x_3, x_4)}{\partial(x_1, x_4)}\right) = 0 \tag{4.27}$$

with the associated Boolean equation

$$[(x_1 \oplus x_2) \wedge v_{03}] \vee [(x_1 \oplus x_3) \wedge v_{05}] \vee [(x_1 \oplus x_4) \wedge v_{09}] = 0 \tag{4.28}$$

can be solved by the PRP:

```
 1   space 32 1                                    39   isc 18 22 23
 2   avar 1                                        40   maxk 23 22 24
 3   v00 v01 v02 v03 v04 v05 v06 v07               41   vtin 1 25
 4   v08 v09 v10 v11 v12 v13 v14 v15               42   v00 v01 v04 v05 v08 v09 v12 v13.
 5   x1 x2 x3 x4.                                   43   vtin 1 26
 6   sbe 1 1                                        44   v02 v03 v06 v07 v10 v11 v14 v15.
 7   (x1#x2)&v03+(x1#x3)&v05+                       45   cco 24 25 26 27
 8   (x1#x4)&v09=0.                                 46   isc 21 27 28
 9   sbe 1 2                                        47   obb 28 28
10   v00=0.                                         48   sbe 1 29.
11   isc 1 2 3                                      49   x3=0.
12   cpl 2 4                                        50   isc 28 29 30
13   isc 1 4 5                                      51   maxk 30 29 31
14   vtin 1 6                                       52   cpl 29 32
15   v01 v02 v03 v04 v05 v06 v07                    53   isc 28 32 33
16   v08 v09 v10 v11 v12 v13 v14 v15.               54   maxk 33 32 34
17   cel 5 6 7 /01 /10                              55   vtin 1 35
18   uni 3 7 8                                      56   v00 v01 v02 v03 v08 v09 v10 v11.
19   obb 8 8                                        57   vtin 1 36
20   sbe 1 9.                                       58   v04 v05 v06 v07 v12 v13 v14 v15.
21   x1=0.                                          59   cco 34 35 36 37
22   isc 8 9 10                                     60   isc 31 37 38
23   maxk 10 9 11                                   61   obb 38 38
24   cpl 9 12                                       62   sbe 1 39.
25   isc 8 12 13                                    63   x4=0.
26   maxk 13 12 14                                  64   isc 38 39 40
27   vtin 1 15                                      65   maxk 40 39 41
28   v00 v02 v04 v06 v08 v10 v12 v14.               66   cpl 39 42
29   vtin 1 16                                      67   isc 38 42 43
30   v01 v03 v05 v07 v09 v11 v13 v15.               68   maxk 43 42 44
31   cco 14 15 16 17                                69   vtin 1 45
32   isc 11 17 18                                   70   v00 v01 v02 v03 v04 v05 v06 v07.
33   obb 18 18                                      71   vtin 1 46
34   sbe 1 19.                                      72   v08 v09 v10 v11 v12 v13 v14 v15.
35   x2=0.                                          73   cco 44 45 46 47
36   isc 18 19 20                                   74   isc 41 47 48
37   maxk 20 19 21                                  75   obb 48 48
38   cpl 19 22
```

and has 32 solution functions which can be expressed by eight ternary vectors of $S(\mathbf{v})$

$$
XBO[48] = S(\mathbf{v}) =
\begin{array}{cccccccccccccccc}
v_{00} & v_{01} & v_{02} & v_{03} & v_{04} & v_{05} & v_{06} & v_{07} & v_{08} & v_{09} & v_{10} & v_{11} & v_{12} & v_{13} & v_{14} & v_{15} \\
\hline
- & 0 & 0 & 0 & 0 & 0 & 0 & 0 & 0 & 0 & 0 & 0 & 0 & 0 & 0 & - \\
- & 0 & 0 & 0 & 0 & 0 & 0 & 1 & 0 & 0 & 0 & 1 & 0 & 1 & 1 & - \\
- & 0 & 0 & 1 & 0 & 1 & 1 & 0 & 0 & 1 & 1 & 0 & 1 & 0 & 0 & - \\
- & 0 & 0 & 1 & 0 & 1 & 1 & 1 & 0 & 1 & 1 & 1 & 1 & 1 & 1 & - \\
- & 1 & 1 & 0 & 1 & 0 & 0 & 0 & 1 & 0 & 0 & 0 & 0 & 0 & 0 & - \\
- & 1 & 1 & 0 & 1 & 0 & 0 & 1 & 1 & 0 & 0 & 1 & 0 & 1 & 1 & - \\
- & 1 & 1 & 1 & 1 & 1 & 1 & 0 & 1 & 1 & 1 & 0 & 1 & 0 & 0 & - \\
- & 1 & 1 & 1 & 1 & 1 & 1 & 1 & 1 & 1 & 1 & 1 & 1 & 1 & 1 & - \\
\end{array}
\qquad (4.29)
$$

with the normal form (3.74).

3.12 The BDE (3.174) has the searched function set as solution. The PRP of Figure 3.8 can be reused after the change of line 8 by:

```
 8                  /u3&w9=1.
```

The set of solutions $S(\mathbf{v}) = XBO[12]$:

$$S(\mathbf{v}) = \begin{array}{c|cccc} & v_0 & v_1 & v_2 & v_3 \\ \hline & 0 & 1 & 1 & 0 \\ & 1 & 0 & 0 & 1 \end{array} \tag{4.30}$$

contains the linear functions:

$$1: \quad f_0(x_i, x_j) = x_i \odot x_j\ , \quad f_1(x_i, x_j) = x_i \oplus x_j \tag{4.31}$$

of one function class.

3.13 Line 9 of the PRP of Figure 3.9 must be replaced by:

$$9 \qquad\qquad z1 \& z2 \& /z3 \& /z4 \& /z5 \& /z6 = 1\ .$$

The BDE (3.258) has 32 solution functions.

3.14 The BDE:

$$q(x_i, x_j) \wedge \max_{x_i} r(x_i, x_j) \wedge \max_{x_j} r(x_i, x_j) = 0 \tag{4.32}$$

has as solution all pairs of functions $\langle q(x_i, x_j),\ r(x_i, x_j)\rangle$ that describe all lattices which include at least one OR-bi-decomposable function $f(x_i, x_j)$. Lines 7 to 17 of the PRP of Figure 3.13 must be replaced by:

```
 7   zr1  zr2 .                    13   _maxk  3  < zr1 >  4
 8   sbe  1  1                     14   sbe  1  5
 9   uq0&zr1&zr2 = 0.              15   zr2 = ur0 + ur2 .
10   sbe  1  2                     16   isc  4  5  6
11   zr1 = ur0 + ur1 .            17   _maxk  6  < zr2 >  7
12   isc  1  2  3
```

to get a PRP that solves this task. The BDE (4.32) has 71 pairs of solution functions which describe 71 lattices of the searched OR-bi-decomposable functions.

Bibliography

J. S. B. Akers, "On a theory of Boolean functions", *Journal of the Society for Industrial and Applied Mathematics (SIAM)*, 1959, Volume 7, No. 4, pp. 487–498. DOI: 10.1137/0107041. 31

D. Bochmann, *Binary Systems - A BOOLEAN Book*, 2008, TUDpress, Verlag der Wissenschaft GmbH, Dresden, Germany, ISBN 978-3940046871. 32

D. Bochmann and Ch. Posthoff, Binäre Dynamische Systeme, 1981, Berlin: Akademie-Verlag, Oldenbourg, ISBN 978-3486250718. 31, 32

M. Davio, J.-P. Deschamps and A. Thayse, *Discrete and Switching Functions*, 1978, McGraw-Hill International, ISBN 978-007015509. 31

D. Bochmann, F. Dresig, and B. Steinbach, "A new decomposition method for multilevel circuit design," *European Conference on Design Automation*, 1991, Amsterdam, Holland, DOI 10.1109/EDAC.1991.206428, pp. 374–377. DOI: 10.1109/EDAC.1991.206428.

D. A. Huffman, Solvability criterion for simultaneous logical equations," *Quarterly Progress Report*, M.I.T. Research Lab. of Electronics, 1958, Volume 1, No. 56, pp. 87–88. 31

A. Mishchenko, B. Steinbach, and M. Perkowski, "An algorithm for bi-decomposition of logic functions," *Proceedings of the 38th Design Automation Conference*, 2001, Las Vegas (Nevada), USA, DOI 10.1109/DAC.2001.156117, pp. 103–108. DOI: 10.1145/378239.378353.

Ch. Posthoff and B. Steinbach, *Logic Functions and Equations - Binary Models for Computer Science*, 2004, Dordrecht, The Netherlands: Springer, ISBN 978-1441952615.
DOI: 10.1007/978-1-4020-2938-7. 31, 110

I. S. Reed, "A class of multiple-error-correcting codes and the decoding scheme," *Transactions of the IRE Professional Group on Information Theory*, 1954, Volume 4, No. 4, DOI 10.1109/TIT.1954.1057465, pp. 38–49. DOI: 10.1109/TIT.1954.1057465. 31

O. S. Rothaus, "On "bent" functions," *Journal of Combinatorial Theory*, Series A, Volume 20, Issue 3, 305, 1976. DOI: 10.1016/0097-3165(76)90024-8. 81

R. Scheuring and H. Wehlan, "On the design of discrete event dynamic systems by means of the Boolean differential calculus," *First IFAC Symposium on Design Methods of Control Systems*, 1991, Pergamon, pp. 723–728.

B. Steinbach, "XBOOLE - a toolbox for modelling, simulation, and analysis of large digital systems," *System Analysis and Modelling Simulation*, 1992, Volume 9, No. 4, pp. 297–312. 31

B. Steinbach, *Lösung binrer Differentialgleichungen und ihre Anwendung auf binäre Systeme*, 1981, Dissertation A (PhD - thesis), TH Karl-Marx-Stadt. 31, 70, 88

B. Steinbach and R. Hilbert, "Schnelle Testsatzgenerierung, gestützt auf den Booleschen Differentialkalkül," *Fehler in Automaten*, 1989, Verlag Technik, Berlin, Germany, ISBN 978-3341006832, pp. 45–90.

B. Steinbach and Ch. Lang. "Exploiting Functional Properties of Boolean Functions for Optimal Multi-Level Design by Bi-Decomposition," *Artificial Intelligence in Logic Design*, 2004, Kluwer Academic Publisher, Dordrecht, The Netherlands, pp. 159–200.
DOI: 10.1023/B:AIRE.0000006606.01771.8f.

B. Steinbach and Ch. Posthoff, "Extended Theory of Boolean Normal Forms," *Proceedings of the 6th Annual Hawaii International Conference on Statistics, Mathematics and Related Fields*, 2007, Honolulu, Hawaii, pp. 1124–1139.

B. Steinbach and Ch. Posthoff, *Logic Functions and Equations - Examples and Exercises*, 2009, Springer Science + Business Media B.V., ISBN 978-1402095948. 31

B. Steinbach and Ch. Posthoff, "Boolean Differential Calculus," *Progress in Application of Boolean Functions* edited by T. Sasao and J. T. Butler, 2010, Morgan & Claypool Publishers, San Rafael, CA, USA , pp. 55–78, 121–126. DOI: 10.2200/S00243ED1V01Y200912DCS026. 107, 108

B. Steinbach and Ch. Posthoff, "Boolean Differential Calculus - Theory and Applications," *Journal of Computational and Theoretical Nanoscience*, 2010, American Scientific Publishers, Valencia, CA, USA, ISSN 1546-1955, DOI 10.1166/jctn.2010.1441, Volume 7, No. 6, pp. 933–981. DOI: 10.1166/jctn.2010.1441.

B. Steinbach and Ch. Posthoff, "XBOOLE and the Education of Engineers," *Computer - Aided Design of Discrete Devices - CAD DD 2007, Proceedings of the Sixth International Conference*, 2007, Minsk, Belarus, Volume 2, pages 14–22.

B. Steinbach and Ch. Posthoff, "Classes of Bent Functions Identified by Specific Normal Forms and Generated Using Boolean Differential Equations", 2011, *Facta Universitatis*, NIŠ, Serbia, Series: Electronics And Energetics, Volume 24, Issue No. 3, ISSN 0353-3670, DOI number 10.2298/FUEE1103357S, pages 357–383. 81, 120, 122, 127

B. Steinbach, F. Schumann, and M. Stöckert, "Functional Decomposition of Speed Optimized Circuits," *Power and Timing Modelling for Performance of Integrated Circuits*, 1993, IT Press, Bruchsal, pp. 65–77.

B. Steinbach and M. Stöckert, "Design of fully testable circuits by functional decomposition and implicit test pattern generation," *Proceedings of the 12th IEEE VLSI Test Symposium*, 1994, Cherry Hill (New Jersey) USA, DOI 10.1109/VTEST.1994.292339, pp. 22–27. DOI: 10.1109/VTEST.1994.292339.

B. Steinbach and A. Wereszczynski, "Synthesis of Multi-Level Circuits Using EXOR-Gates," *IFIP WG 10.5 - Workshop on Applications of the Reed-Muller Expansion in Circuit Design*, 1995, Chiba - Makuhari, Japan, pp. 161–168.

B. Steinbach and Z. Zhang. "Synthesis for full testability of large partitioned combinational circuits," *Boolesche Probleme, Proceedings of the 2. Workshop*, 1996, Freiberg, Germany, pp. 31–38.

B. Steinbach and Z. Zhang. "Synthesis for Full Testability of Partitioned Combinational Circuits Using Boolean Differential Calculus," *IWLS'97 - Synthesis in the Sierra*, 1997, Granlibakken Resort - Tahoe City, CA - US, pp. 1–4.

A. Thayse, *Boolean Calculus Of Differences*, 1981, Berlin, Springer, ISBN 978-3540102861. DOI: 10.1007/3-540-10286-8. 31

S. N. Yanushkevich, *Logic Differential Calculus in Multi-Valued Logic Design*, 1998, PhD thesis, Tech. University of Szczecin, Szczecin, Poland. 32

Authors' Biographies

BERND STEINBACH

From 1973–1977, Bernd Steinbach studied Information Technology at the University of Technology in Chemnitz (Germany) and graduated with an M.Sc. in 1977. He graduated with a Ph.D. and with a Dr. sc. techn. (Doctor scientiae technicarum) for his second doctoral thesis from the Faculty of Electrical Engineering of the Chemnitz University of Technology in 1981 and 1984, respectively. In 1991 Steinbach obtained the habilitation (Dr.-Ing. habil.) from the same Faculty. Topics of his theses involved Boolean equations, Boolean differential equations, and their application in the field of circuit design using efficient algorithms and data structures on computers.

Steinbach worked in industry as an electrician, where he had tested professional controlling systems at the Niles Company. After his studies he taught as Assistant Lecturer at the Department of Information Technology of the Chemnitz University of Technology. As a research engineer he developed programs for test pattern generation for computer circuits at the company Robotron. He later returned to the Department of Information Technology of the Chemnitz University of Technology as Associate Professor for design automation in logic design.

Since 1992 he has been a Full Professor of Computer Science/Software Engineering and Programming at the Freiberg University of Mining and Technology, Department of Computer Science. He has served as Head of the Department of Computer Science and Vice-Dean of the Faculty of Mathematics and Computer Science.

His research areas include logic functions and equations and their application in many fields, such as Artificial Intelligence, UML-based testing of software, UML-based hardware/software co-design. He is the head of a group that developed the XBOOLE software system. He published three books in logic synthesis. The first one, co-authored by D. Bochmann, covers Logic Design using XBOOLE (in German). The following two, co-authored by Christian Posthoff, are *Logic Functions and Equations–Binary Models for Computer Science* and *Logic Functions and Equations–Examples and Exercises*, Springer 2004, and 2009, respectively. He published more than 200 chapters in books, complete issues of journals, and papers in journals and proceedings.

He has served as Program Chairman for the IEEE International Symposium on Multiple-Valued Logic (ISMVL), and as guest editor of the *Journal of Multiple-Valued Logic and Soft Computing*. He is the initiator and general chair of a biennial series of International Workshops on Boolean Problems (IWSBP) which started in 1994, with 10 workshops until now.

He received the Barkhausen Award from the University of Technology Dresden in 1983.

CHRISTIAN POSTHOFF

From 1963–1968, Christian Posthoff studied Mathematics at the University of Leipzig. His thesis was titled: "Axiomatic Description of a Finite Class Calculus" (Prof. Dr. Klaua).

From 1968–1972, he worked as a programmer and in the field of Operations Research; simultaneously, he did his Ph.D. in 1975 with the thesis "Application of Mathematical Methods in Communicative Psychotherapy."

In 1972, he joined the Department of Information Technology at the Chemnitz University of Technology; up to 1983, his research activities concentrated on logic design. His cooperation with B. Steinbach goes back to these days. Important results have been algorithms and programs for solving Boolean equations with a high number of variables and the Boolean Differential Calculus for the analytical treatment of different problems in the field of logic design. These results have been collected in a monograph *Binary Dynamic Systems* published simultaneously in Akademie-Verlag Berlin, Oldenbourg-Verlag Munich-Vienna and in the Soviet-Union, and allowed the habilitation (Dr.-Ing. habil.) at the Faculty of Electrical Engineering in 1979 and the promotion to Associate Professor. He wrote two textbooks at this time, aimed at a higher level in the theoretical and mathematical training of graduate engineers of information technology. About 1976, he started research activities in Artificial Intelligence.

In 1983, Posthoff started as Full Professor of Computer Science in the Department of Computer Science at the same university, with the aim of starting the program in Computer Science in 1984. In 1984, he became Head of the Institute of Theoretical Computer Science and Artificial Intelligence and Research Director of the Department of Computer Science. An independent direction of research activities within AI, investigations of computer chess and other strategic games, arose from his love for chess. His research activities concentrated on the application of fuzzy logic for the modeling of human-like "thinking" methods, the learning from examples, the construction of intelligent tutoring systems, the parallelization of inference mechanisms, systems of diagnosis and configuration. In cooperation with colleagues from different areas of mechanical engineering and medicine, he has been supervising the construction of several expert systems. He received the Scientific Award of the Chemnitz University of Technology four times.

In 1994, he moved to the Chair of Computer Science at the University of The West Indies, St. Augustine, Trinidad & Tobago. From 1996–2002 he was Head of the Department of Mathematics & Computer Science. His main focus was the development of Computer Science education at the undergraduate and graduate level to attain international standard. In 2001, he received the Vice-Chancellor's Award of Excellence.

He is the author or co-author of 15 books and many publications in journals and conference proceedings.

Index

algorithm
    separation of function classes, 73
    separation of functions, 93

bent functions
    4 variables, all, 120
    4 variables, most complex, 110
bi-decomposition
    AND (x_i, x_j), 109
    AND $(\mathbf{x}_a, \mathbf{x}_b)$, 109
    EXOR (x_i, x_j), 109
    EXOR $(\mathbf{x}_a, \mathbf{x}_b)$, 109
    OR (x_i, x_j), 109
    OR $(\mathbf{x}_a, \mathbf{x}_b)$, 109
Boolean Differential Calculus, 31
Boolean differential equation
    all derivative operations, 105
    derivatives in all directions of change, 56
    each set of solution functions, 86
    function classes, 56
    most general, 122
    single simple derivative, 49
    single vectorial derivative, 52
Boolean Differential Equation, 49
Boolean equation, 9
    inequality, 9
    solution, 9
    solution under constraints, 13
    solution with regard to variables, 10
    systems of equations, 9
Boolean function

monotonously falling, 106
monotonously rising, 105
symmetric, 100
Boolean space, 20

derivative operation
    Δ - operation, 43
    m-fold derivative, 41
    m-fold maximum, 43
    m-fold minimum, 42
    simple derivative, 32
    simple maximum, 33, 34
    simple minimum, 33
    vectorial derivative, 37
    vectorial maximum, 37
    vectorial minimum, 37

exercises, 29, 48, 126
    solutions, 129
extension sdt, 17

formula, 4
    equivalent formula, 6
    interpretation, 5
    logical law, 6
    tautology, 6

hazard
    functional static 0-hazard, 107
    functional static 1-hazard, 108

integrability condition, 50, 52, 54, 55

Karnaugh map, 18

lattice, 123
logic equation, 9
    inequality, 9
    solution, 9
    solution under constraints, 13
    solution with regard to variables, 10
    systems of equations, 9

supplementary function
    d2v($SLS(\mathbf{u})$), 73
    d2v($SLS(\mathbf{u}), \mathbf{x}$), 90
    epv($S(\mathbf{v}), i$), 73
    epv($S(\mathbf{v}, \mathbf{x}), i$), 92

transformation as preprocessing step
    multiple step transformation, 120
    single step transformation, 117
TVL, 18

XBOOLE Monitor, 16
XBOOLE-Monitor
    4-fold View, 18, 22
    Append Ternary Vector(s)..., 22
    Context Help, 20
    Create TVL..., 20, 21
    Define Space..., 20
    Delete Protocol, 26
    Derivatives, 24
    Execute PRP... , 26
    Extended Operations, 24
    General, 21
    Help Topics, 20, 23
    Index of Commands (ordered by topics), 24
    List of the Commands, 24
    Matrices, 24

Object Management, 24
Objects, 20
Open PRP..., 26
PRP, 18
Save protocol as PRP..., 26
Sets, 24
Single View, 18, 22
Spaces/Objects, 18
Summary of the toolbars, 21
Survey of the Toolbars, 21
sdt-file, 17
vtin, 25
button Create, 22
button K, 18
button Single Step, 26
button T, 18
command avar, 26
command obb, 27
command language, 23
command line, 19, 23
edit mode, 23
help information, 20
menu '?', 23
menu bar, 17
orthogonal disjunctive/antivalence (ODA) form, 18
problem program, 18
problem program (PRP), 25
protocol, 18
toolbar, 18, 21
toolbar Objects, 21
toolbar icon, 21
toolbars Derivatives, Matrices, Sets and Extras, 21
TVL mode, 23
VT, 18
XBOOLE help system, 19

Printed in the United States
by Baker & Taylor Publisher Services